OF ICE AND MEN

OF ICE AND MEN

HOW WE'VE USED COLD TO TRANSFORM HUMANITY

FRED HOGGE

PEGASUS BOOKS
NEW YORK LONDON

OF ICE AND MEN

Pegasus Books, Ltd.
148 West 37th Street, 13th Floor
New York, NY 10018

Copyright © 2022 by Fred Hogge

First Pegasus Books cloth edition December 2022

Interior design by Maria Fernandez

ISBN: 978-1-63936-183-0

10 9 8 7 6 5 4 3 2 1

Printed in the United States of America
Distributed by Simon & Schuster
www.pegasusbooks.com

For my mother,

who very much wanted to read it.

And for Karen,

who always believed I could actually write it.

CONTENTS

INTRODUCTION

Once upon a time in the fertile lands of southern Iraq, in those distant days when gods still walked among us, the goddess Inanna sailed out to see her father, Enki, the god of wisdom.* He lived on an island called Eridu, the very birthplace of all creation, where land first rose from the primordial sea at the dawn of time. He greeted her boat, welcomed her ashore and to his shrine, and together they drank beer late into the night. And in the morning, when she left, she took with her his gift of *me*†— the divine powers, the gift of civilization itself.

When Enki woke and found it gone, he sent word to Inanna, demanding its return. She replied, "How could my father change his mind and break the promise he made me

---

* Please note that this is in no way a direct translation of the story, which you can find in titles like *Myths from Mesopotamia: Creation, The Flood, Gilgamesh, and Others,* translated by Stephanie Dalley.

† Sometimes rendered in English as "*meh*" which, with its modern connotations, feels somehow inappropriate given its awe-inspiring nature.

with this gift?" Rather than yielding it, she chose instead to bestow it upon mankind. But this gift, as wonderful as at first it seemed, came at a price. For it held not only all the pleasures of a civilized life, of art and fashion, sex and music, it held their opposites as well. The art of kindness lay balanced with the art of might. That of straight-forwardness came with its counterpart deceit, pity with terror, justice with strife, peace with war. If you accept it, Enki warned us mortals, you must take all of it, and once you take it, you can never give it back.

You may be wondering, "What the hell does this original cautionary tale have to do with ice?" Simply this: "Ice," Allie Fox argues, in the film *The Mosquito Coast*, "*is* civilization." And during the course of just over two hundred years, humanity has taken it from a rare, luxury commodity and used it to fundamentally transform our species in ways to which we rarely pay attention and with consequences we are only just beginning to fully understand.

In this book, I've set out to gather together the stories that make up that transformation, tales of pioneers, inno-vators, and stubborn entrepreneurs, to try to explore the often unintended consequences of their actions.

Consequences are, of course, inevitable. But in a his-torical context, they sadly lack the comforting symmetry of those we see in our physical world as laid out by Isaac Newton in his Third Law of Motion. Their reactions are rarely equal or opposite. Instead, they veer wildly from the positive to the tangential to those which seem out of all proportion to the event that triggered them.*

The historical citizens of Uruk, alleged descendants of those mythological forebears to whom Inanna gave so much, could never have imagined the consequences of their creation: the world's first city.† Its people invented writing and the first laws. They devised units of time and mapped the stars. They even created, as far-fetched as it might seem, the first ice stores in what is now desert.

In its first two hundred years, their city grew tenfold. The new concept of civic life allowed their population to explode to such an extent that, in time, their lands could no longer bear the burden of the demands for food and water placed upon it.

Uruk's once-fertile plains in time became a barren, wind-scarred landscape. And its people could no more have foreseen this desertification nor the abandonment of their city than their scribes have envisaged the works of Shakespeare or their soldiers the horrors wrought on their lands by the wars of recent years.

Much like the people of Uruk, we live in a time of brilliant innovation which has brought about vast increases in our numbers. Between 1950 and 2020, the population of

---

* The assassination of one posh Austrian leading to the slaughter of the First World War being a case in point.

† It was founded, according to the Sumerian King List, by King Enmerkar in or around 4500 B.C.E.

the world's largest city, Tokyo, has tripled to 34.4 million people. In the same period, cities like Houston, Texas, and Sydney, Australia, have broadly kept pace, tripling in size. Meanwhile, Sao Paolo, Brazil, has quintupled in size in this timeframe, while Lagos, Nigeria, now home to 14.3 million, has swollen by a staggering forty-three times.*

In fact, the global population as a whole has more than tripled since 1950, from 2.5 to 7.7 billion people.

Much of this has to do with the obvious strides humanity has taken in its ability to provide itself with food and healthcare. But hidden away behind the food we eat and the medicines we take is an infrastructure based on ice, air cooling, and refrigeration, none of which we had as little as two hundred years ago, and certainly not on the industrial scale of today.

We have placed ice deeply within the vital systems of our global society to such an extent that it has supercharged our species in ways, like the scribes of Uruk, we can scarcely imagine. It cools our drinks, it stores our food, it cools our houses. We use it in sport, in medicine, in technological innovation. We even use the detection of its presence to explore the universe. Our use of it is changing us as once did its elemental opposite: fire.

◆

---

* One thing these cities have in common? Air conditioning.

When I began to think about this book, I had one humble and unanswerable question (though we'll get as close as I can to one in chapter three): When did ice first appear in the cocktail? Historical cocktail recipe books take its presence for granted. But, once there, its ubiquity renders it unremarkable.

When you start looking for it, you begin to see ice and its related technologies everywhere, and equally unremarked upon.

And this begs a question: If the fictional Allie Fox was right and ice is indeed civilization, what might be the price for it that Enki demands we pay?

OF ICE AND MEN

1

Life in the Freezer

Ice is a serial killer. It's not that it wants you dead,* but, given the right conditions and enough time, it will end you.

This is how.

When the cold first takes hold of you, your body starts to shiver. Anyone who has lived through a harsh winter will have felt this, safe in the knowledge that they can go back inside in a little while, and all will be well. But each shiver is an involuntary action designed to generate heat, burning energy in the short term to keep you warm.

Alongside the shivering, you'll notice your fingers and toes numbing, again, involuntarily. Your hypothalamus has taken charge, drawing blood away from your peripheries to hold as much warmth within your core as possible.

---

* It is, after all, inanimate.

Think of this as the cold trying to buy you a drink at the bar. It hasn't yet offered you a lift home.

As your temperature continues to drop, your shivering increases. And you won't notice how gradually your coordination begins to drift, nor that your speech has become slurred. You will not grasp that most of the things you say will have little bearing on your situation.

You are now suffering from moderate hypothermia, and it is messing with your consciousness. When you get a little colder, you are going to start feeling warmer, as though your body's survival strategies are working. In reality, your blood vessels are widening to allow more blood and precious heat to your vital organs—the opposite process to the finger-numbing you felt earlier on. And it is so effective that some people, about a fifth to half of hypothermia victims, will now undress to cool down, a phenomenon known as Paradoxical Undressing.

If help comes now, you will be fine.

It's when you stop shivering that you really need to worry, though, and thanks to your drifting consciousness, you won't be able to. Your body is now burning so much energy as it tries to keep you warm that it must stop . . . to keep you warm.

The cold has you now. Your pupils dilate, your pulse rate drops, and you feel . . . well . . . pretty good, actually. Survivors will tell you that this stage of hypothermia produces a trippy high. But as good as you feel inside, outwardly you look dead. And without rescue, you soon will be.

Your heart rate, blood pressure, and respiration have all slowed. You have one little piece of hardwired, involuntary behavior left on your side: Terminal Burrowing. You will try to squeeze yourself into the smallest space possible, like behind a wardrobe or under a bed, in the unconscious hope that the little body heat you have left might warm the space, and thus you.

As your body temperature drops below 20°Celsius (68°Fahrenheit), you will display no vital signs at all.

You are now basically dead. Another victim of the cold.

And yet, there are plenty of examples of hypothermia victims being saved against all the odds, including the one that opens the first chapter of Atul Gawande's book *The Checklist Manifesto*. It's an account of a three-year-old Austrian girl who fell through the ice on a fishpond in the one brief moment her parents' backs were turned. She was in the water for half an hour and, by the time she was plugged into life support machines at the nearest hospital, she had been dead for ninety minutes. Two weeks later, she was back at home after making a full recovery.

To revive someone like this, you need a lot of things to go right in the right order. And that's before we even consider this little girl's particular advantage: there was help immediately at hand.

Imagine, then, stepping outside, unprepared, in the High Arctic. It is noon in mid-December in Utqiagvik, previously Barrow, Alaska, the most northerly town in the

United States. Snow is heavy on the ground. The roads are gritted, the streetlamps are on. The sky is a deep shade of luminescent indigo, the color it turns everywhere else in the world when the sun has just dipped beneath the horizon.

Here, it has not even climbed above it. This is as bright as the day will get. Without the sun's warmth, an average high temperature might read somewhere about -19°C (-2°F, which sounds a lot less threatening). But, as you close the door to your heated room behind you, the cold will grab onto you like quicksand. You will hear your every outward breath fall as ice crystals on the ground, making a sound some call the whisper of the stars. Your only source of warmth will be from your own body. And any breath of wind will find its way through the merest gap in your clothes to nip at your skin. Anything exposed, an earlobe, a nose tip, is now at risk.

You are on borrowed time in an environment that screams at you in every way: you do not belong here.

Of course, you're still in town, and you can go inside. So let us picture something more extreme: instead of simply walking out your door, you're in an airplane, minding your business, flying along through the fiercely changeable skies of northern Alaska. Perhaps you were bound for Utqiagvik before trouble plucked you from the air. The controls or the fuel line freezes up. You're out of control in the bottom of nowhere, but somehow you land.

How, you cannot be sure. But you're on the ground, alive, and you're upside down.

You. Are. Screwed. And you know it. And there is nothing you can do about it.

Unless someone just happens to be passing.

As improbable as this sounds, it is exactly what happened to Peter Merry.*

Merry was an experienced bush pilot who spent the bulk of his career flying across Alaska in everything from single-engined Norsemen to 737s over the course of forty years. He knew the Arctic and he knew the risks. He knew how the cold could pluck a plane from the sky as though on a whim.

"It wasn't an easy business," he'd later say of it. "When I went up to Port Barrow† there were nine of us . . . And three years later, there were four of us."‡

Had it not been for a man named Harry Brower Sr., there would have been one less.

According to his friend Sam P. Hopson, Brower seemed to have an intuition for finding and helping people in trouble, honed over a lifetime spent in the High Arctic. Brower's niece and sister-in-law, Mable Hopson, described it as an instinct. The writer and academic Karen Brewster,

---

* *The Whales They Give Themselves*, p.179.

† Utqiagvik.

‡ Interview with *Project Jukebox, University of Alaska Fairbanks Oral History Program*, conducted by Bill Schneider and David Krupa, October 29, 2002.

who recorded conversations with Harry for her book, *The Whales They Give Themselves*, wrote about how his mixed-race upbringing, half white, half Iñupiat, opened before him a world of possibilities.

"While a devout Christian, Kupaaq* grew up hearing stories that reinforced that the world was alive with humans and non-humans, natural and supernatural interactions that should be paid attention to." His was a world where the notion of a woman turning herself into a bear was entirely normal. "It helped him," Brewster wrote, "understand seemingly inexplicable sensations he had as an adult, like knowing that a whale was about to come up or sensing when people were in trouble."

His indigenous Iñupiaq mother's culture is one where shamans were known to transform into animals, to bless people with successful hunts, or to curse them with bad fortune. In short, a world that science finds hard to understand.

And yet, Harry Brower, the man with a sixth sense who was able to describe the plight of a baby whale accidentally hunted by his son while lying in an Anchorage hospital bed,[†] was also one of the best friends of Arctic scientists, and the man who almost singlehandedly transformed the world's understanding of bowhead whales.

Veterinary scientist Tom Albert described Brower as one of the best teachers he ever had. It was to Harry he

---

*   Brower's Iñupiaq name.

†   *The Whales They Give Themselves*, pages 156–160.

would turn for information, be it about whales or all manner of Arctic fauna. It was Harry who persuaded the other subsistence whaling captains to allow the scientists to take specimens and measurements from the whales they caught. It was Harry who showed the International Whaling Commission that there were many more bowheads out there than scientists believed, and thus saved a vital part of his Indigenous culture.* "Harry's the guy who was in behind the scenes who had the biggest impact of any single person," Albert said.

Brower bridged the gap between ancient and modern knowledge, gleaned from Iñupiaq elders and from his mother, who hunted for the family. Those lessons were built around practice and storytelling.

As Albert would later write: "From having spoken to many hunters and from extended conversations with Harry Brower Sr., it became clear that there was a very specific body of knowledge regarding the bowhead whale that was held by these people. This knowledge had been handed down from fathers to sons for generations, it was tested over many years, it definitely had survival value, and in view of this the designation 'Traditional Knowledge' seemed appropriate." And he would conclude that, "The success of this program is strong evidence that scientists and other technical people should carefully consider the traditional knowledge held by local people."†

---

* Subsistence whaling.

† In the Journal of the Arctic Institute of North America, 2000.

It may seem surprising that Albert felt the need to state this. You'd think that, if you're about to set out into a wilderness about which you know next to nothing, it might be a good idea to consult an expert. But, in the history of polar science, taking local advice has always been the exception rather than the rule.

This is all the more baffling when one realizes that the polar explorers who set off to chart the Arctic Ocean's coasts in the early nineteenth century knew less about their destination than Apollo space engineers knew of the lunar surface before they landed Armstrong and Aldrin upon it. They weren't even sure that the lunar surface could or would support the weight of the *Eagle* lander, with some fearing it might be swallowed by dust should they touch down successfully. In 1800, there was a widely held belief that, if one could sail beyond the pack ice found between 70° and 80° North, one would discover open ocean.

Among the proponents of an open polar sea was William Scoresby, a whaling captain who was about as close as you could get to a European Arctic expert. Scoresby used whales to make his case, arguing that they couldn't breathe if all there was atop the world was ice. Moreover, he wondered, how did some of them end up with stone lances, "a kind of weapon used by no nation now known," embedded in their blubber?

As preposterous as it seems to us today, the idea proved influential, particularly upon John Barrow, who became Second Secretary of the Admiralty in 1804, and would be one of the key architects of British polar exploration in the 1800s. For it held within it a beguiling promise: that of a Northwest Passage to the Orient.

Barrow was far from the first to be seduced by the notion. As Scoresby wrote in 1820, two years after Barrow's first mission, commanded by John Ross, had left Great Britain, "There have only been three or four intervals of more than fifteen years in which no expedition was sent out in search of one or other of the supposed passages from the year 1500."*

In fact, Henry VII of England commissioned the first such mission in 1497, sending John Cabot out to find a route around the north. He was followed by Estêvão Gomes, under the orders of Holy Roman Emperor Charles V in 1524, and Martin Frobisher in 1576. Meanwhile, from the west, Francisco de Ulloa was dispatched from Mexico out into the Pacific by Hernán Cortés to find the so-called Strait of Anián, alleged to connect the Pacific to the Gulf of St. Lawrence, in 1539—an expedition so successful at exploration and mapmaking that it led to the misconception that California was an island, a notion not dispelled until the eighteenth century.

The missions sent forth by Barrow and his successors, funded by the Royal Navy and with reward incentives offered by the Royal Society, would prove just as successful.

---

* The "supposed passages" he refers to here were hoped to run around the top of Canada and Alaska in the west, and around the north of Russia and Scandinavia in the east.

The most famous—the fabled Franklin Expedition*—
would vanish completely, leaving behind only a mystery
that would last more than 150 years†.

Between Ross's first expedition and Franklin's last,
the Royal Navy would send another seven missions to
explore and map the openings to the Northwest Passage.‡
Franklin himself served on two overland trips, the first
of which almost ended in disaster. He lost eleven of his
twenty men to starvation or exhaustion, one murder,
and claims of cannibalism. Not that this would impede
his career. Instead, it brought him national fame as "the
man who ate his boots," thanks to the fact he and his
men were so starved that these and lichen were all that
were left on the menu.

It is hardly surprising then, that when his final
expedition set sail from Greenhithe on May 19, 1845,
it was arguably the best supplied to have put forth in
British history, thanks to lessons learned. There was
food aboard to last three years, longer if they chose to
ration it. They had a library of 1,200 books to help pass
the time when they inevitably became marooned in the

---

*   Named for its commander, Sir John Franklin.

† As we get stuck into the story of Franklin and his colleagues, I wish to
acknowledge my debt to Michael Palin's extraordinary book, *Erebus:
The Story of A Ship*, published in 2018, which has provided me with
most of the expedition's correspondence that I have quoted below.

‡ They would send seven to Antarctica, too. After all, you have to
find something to occupy a navy when you aren't fighting a war,
and to justify the expense of keeping it on.

ice for a long Arctic winter. They had a brick furnace-fired heating system. Both *Erebus* and *Terror* were fitted with steam engines, among the first such ships to be thusly adapted, complete with decouplable propellors to reduce drag and demountable funnels. In addition to the livestock loaded aboard for food, they had a dog, known as Old Nep (short for Neptune), and a monkey called Jacko.* They even had a rubber dinghy.

But for all this provisioning and a full twenty-seven years of recent British naval experience in Arctic and Antarctic waters, their equipment can only be described as woefully insufficient.

The era's Navy-issue cold weather clothing included such garments as boxcloth jackets and trousers, water boots, pairs of hose, woollen neck scarves, and "Welsh wigs," their version of a beanie hat.

While woollens were all very well on board, they have a fundamental problem out on the Arctic tundra. They absorb the sweat from any exertion. As soon as your physical activity ends, that sweat will freeze in the fibers. Suddenly, your very clothes become your enemy.

With their traditional seal skins, this is not an issue for the Inuit.

Franklin's orders from the Admiralty stated that, if he and his crew had to spend the winter in the Arctic,

---

* While Old Nep was, according to Lt. James Fairholme in one of his final letters home, "the most loveable dog I ever knew," Jacko was "a dreadful thief" and "the annoyance and pest of the whole ship."

he should find his ships a safe harbor and take care of his men's health. They also said that, should they meet any Esquimeaux [*sic*], "you are to endeavour by every means in your power to cultivate a friendship with them, by making presents of such articles as you may be supplied with, and what may be useful or agreeable to them; you will, however, take care not to suffer yourself to be surprised by them but use every precaution, and be constantly on your guard against any hostility."

No thought is given to trying to learn from their expertise for survival in the far north.

That being said, when the expedition made its first contact with the Inuit at the Whale Fish Islands, just west of Greenland, a couple of the officers, Henry Goodsir, the assistant surgeon, and James Fairholme, third lieutenant,* did take the time to visit a local settlement and begin compiling a dictionary of Inuit words that might prove useful. Fairholme and James Fitzjames, the *Erebus*'s second in command, also took the chance to try out Inuit kayaks.

Here the crews would complete the revictualling of *Erebus* and *Terror* with stores from their supply ship, the *Baretto Junior*, which, when finished, would sail back to London with news of the voyage thus far.

From these last letters home from the officers and men, we can conjure something of the atmosphere aboard. While many complained of the vicious Greenland mosquitoes,

---

* Both of the *Erebus* crew.

with Ice Master James Reid complaining, "My face and hands are all swollen with their bites," Fitzjames wrote, "You have no conception of how happy we are."

But this stands in stark contrast to the letters of Franklin himself and those of his second in command and captain of the *Terror*, Francis Crozier.

Franklin, who seems so stiff, so tired, so unvital in the series of daguerreotype portraits of the *Erebus* officers shot by the photographer William Beard and commissioned by Franklin's wife, Lady Jane, shortly before his departure, almost seems to be seeking reassurance that he is up to his command. Clearly, he is aware that he was not the first choice for the mission—the Admiralty preferred his friend James Clark Ross, another naval officer who had endured the Arctic six times before. Ross had only returned from the Antarctic just two years earlier, leading an expedition made up again of the ships *Erebus* and *Terror*, and had ruled himself out. Recently married, he had made promises to his wife and father-in-law that he would not return to polar waters, and had recommended Franklin for the assignment.

Crozier, who had captained the *Terror* on that same Antarctic voyage and was another experienced hand, wrote to Ross: "What I fear from being so late [into the Arctic Circle] is that we will have no time to look around and judge for ourselves, but blunder into the ice and make a second 1824 of it." He is referring to a previous expedition, led by William Parry, on which he served as a midshipman, which was caught in the ice for the entire winter. "James," he goes on, "I wish you were here." He continues, "James, dear, I am sadly alone, not a soul have I on either ship that I can talk to." The prospect of finding himself iced in with such (to him) disagreeable company must have been hard to bear.

Still, it was felt that lessons had been learnt from the Parry expedition and others previous, hence the vast provisioning, the library, and an almost-contemporary focus on the physical and mental wellbeing of the men. Any foreboding put down upon paper had to be set aside. The mission was clear. And on July 12, with the provisioning complete, Lieutenant Griffiths of the supply ship *Baretto Junior*, after a lunch of cured beef aboard the *Erebus*, bade goodbye to Franklin and his officers, and sailed for England that afternoon.

Over the remaining weeks of July, Franklin's expedition was seen by three whaling ships in the northern waters of Baffin Bay. On July 26, in what is considered to be last contact, the *Erebus* and the *Terror* rendezvoused with the whaler *Prince of Wales* almost 350 nautical miles from their provisioning anchorage at the Whale Fish Islands, and invited her Captain Dannett to dinner. With the weather in his favor, he had to decline and ultimately sailed on.

No European would see them alive again.

No one knows for sure exactly what happened next. The only information ever found that had been left deliberately by the crew—the so-called Victory Point Note, discovered in 1859, some eleven years after any searches for them began—contains two separate messages.

The first, dated May 28, 1847, says:

> H.M.S ships 'Erebus' and 'Terror' wintered in the Ice in lat. 70 05' N., long. 98 23' W. Having wintered in 1846–7 at Beechey Island[a], in lat. 74 43' 28" N., long. 91 39' 15" W., after having ascended Wellington Channel to lat. 77°, and returned by the west side of Cornwallis Island. Sir John Franklin commanding the expedition. All well.

> Party consisting of 2 officers and 6 men left the ships on Monday 24 May, 1847.

> (Signed) GM. GORE, Lieut.
> (Signed) CHAS. F. DES VOEUX, Mate.

The second, on the same piece of paper, is more desperate.

> [25th April 1848]* H.M. ships 'Terror' and 'Erebus' were deserted on 22 April, 5 leagues

---

* I have the dates written in the Victory Point Note as written by the officers in question. For some reason I have never been able to fathom, our two styles of written English, British and American, present date and month oppositely. Elsewhere, dates are presented in the American style.

N.N.W. of this, [hav]ing been beset since 12th September 1846. The officers and crews, consisting of 105 souls, under the command [of Cap]tain F.R.M. Crozier, landed here in lat. 69° 37' 42" N., long. 98° 41' W. [This p]aper was found by Lt. Irving under the cairn supposed to have been built by Sir James Ross in 1831–4 miles to the Northward—where it had been deposited by the late Commander Gore in May June 1847. Sir James Ross' pillar has not however been found and the paper has been transferred to this position which is that in which Sir J. Ross' pillar was erected—Sir John Franklin died on 11th June 1847; and the total loss by deaths in the expedition has been to this date 9 officers and 15 men.

(Signed) JAMES FITZJAMES, Captain H.M.S. Erebus.

(Signed) F.R.M. CROZIER, Captain & Senior Offr. and start on tomorrow, 26th, for Back's Fish River.

"Never get off the boat." So says Captain Willard in the film *Apocalypse Now*. Here in the High Arctic, the ships meant warmth, shelter, food. The decision to abandon them feels almost like a rejection of life itself. Not least because the Admiralty did not train expedition personnel to live off the land—they were supposed to have everything they needed with them on board the ships. As John Robertson, the ship's surgeon on the aborted Ross/McClintock rescue mission, noted later, they were "not visited by either deer, hare or grouse, nor were we able to provide a single fish."

These were the exact same conditions the Franklin survivors had to endure.

One had to feel for Crozier now. He had 104 men left under his command. Their lives depended on his decisions. Like Shackleton in Antarctica in 1915, as his ship *Endurance* was crushed by the ice, he must have realized that they had no other choice but to find another way out of their predicament. On foot.

But the men of the Franklin Expedition were simply not equipped nor trained to walk out of the Arctic. And, despite their previous contact with the Inuit, they had no local knowledge to assist them.

They were doomed. And Crozier likely knew it.

It is a terrible and tragic coincidence that the first three rescue missions were sent out that same year, 1848, only to be turned back by the brutal Arctic weather that had also done in Franklin. Perhaps, with more clement conditions, they might have found survivors. As it turned out, the fate of the Franklin Expedition remains in large part a mystery despite thirty-six search and rescue missions dispatched over the next decade and the world's subsequent fascination with finding an answer to their fates.

No trace of the expedition would be found until 1850 when the crew of HMS *Assistance* found "fragments of naval stores, portions of ragged clothing [and] preserved meat tins" at Cape Riley, on the southwestern tip of Devon Island.* Meanwhile, HMS *Pioneer's* shore team found a cairn on Beechey Island, but nothing lay beneath it. And a third ship's crew, that of the *Lady Franklin*, discovered three graves on Beechey Island: John Torrington, leading stoker aboard the *Terror*, John Hartnell, able seaman on the *Erebus*, and William Braine, one of the *Erebus's* marines, all of whom had died in 1846.

Further along the coast, a search party under the command of Lieutenant Osborn found sledge runners and a second cairn. This one stood some seven feet high, covering food tins . . . and nothing else. Osborn would write: "Everyone felt that there was something so inexplicable in the non-discovery of any record, some written evidence of the intentions of Franklin and Crozier on leaving this spot."

It must have seemed all the more baffling because Franklin had been explicitly advised to do just that should they be forced off their ships. Franklin and the remainder of his men had seemingly vanished from the face of the earth. Would that there were someone to ask who might know what had happened to them . . .

---

* As recorded by Captain Erasmus in his log.

There was. And he would be found by a Scottish explorer, John Rae.

Rae, the only significant polar explorer of the era never to receive a knighthood, was both a surgeon and a trained surveyor. Arriving in the Arctic for the first time in 1845, he realized quickly that, if you want to get by, you need to listen to those who know how. Like the Norwegian explorers Amundsen and Nansen,* he respected and paid attention to the experts on living in a frozen world: the Inuit. From them he learned how to survive the ice, how to shelter, what to eat, and how to dress. He would, in the course of his career, map some 1,750 miles of unexplored ground with the loss of just one man.

It is hardly surprising that he was chosen to colead the first overland expedition, along with Sir John Richardson, to find news of Franklin in 1848. It was to prove a fruitless mission. But Rae stayed in the Arctic, mapping territory and searching for news. It would take him six years to find it.

On April 21, 1854, he met two Inuit men who would help him uncover the terrible story of the Franklin Expedition's survivors. "Have you seen any foreigners?" Rae asked. One of the men, named Inuk-pukiijuk, said that he hadn't, but he had heard stories of a group who had.

---

* The first men to navigate the Northwest Passage and to cross Greenland respectively.

Rae now spent the summer gathering as much testimony from the Inuit as possible and buying from them artifacts they had gathered from the expedition. By the time he returned to Repulse Bay at the end of July, he had enough information to write a harrowing report.

According to witnesses, four years previously, in 1850, a party of Inuit met about forty white men traveling south near the north shore of King William's Island. They were dragging sledges, one of which had a boat upon it, and were hoping to find a place where they might be able to hunt some deer. They were thin and weak. Their leader, whom the Inuit called Aglooka, was described as tall, broad, and middle-aged. It is now believed this was Francis Crozier. The Inuit sold them some seal meat, and the two groups went their separate ways.

Towards the end of that same season, the returning Inuit group found thirty bodies on the mainland near the Back's Great Fish River. On a nearby island, they found five more. Some had been buried, and some were found under the boat, which had been overturned for shelter. One was found with his telescope still slung about his shoulders and his double-barrelled gun beneath him.

But that was not all.

In his report to the Admiralty, Rae wrote: "From the mutilated state of many of the corpses and the contents of the kettles, it is evident that our wretched countrymen had been driven to the last resource—cannibalism—as a means of prolonging their existence."

Rae wrote a second report to Archibald Barclay, the secretary to the governor and committee of the Hudson Bay Company in London, confirming that, while he was able to crosscheck the story with other Inuit, he was unable to find the bodies himself. "In extenuation of my failure may I mention that I was met by an accumulation of obstacles beyond the usual ones of storms and rough ice, which my former experience in Arctic traveling had not let me to anticipate."

We can assume, given his credentials, that if Rae couldn't do it, nobody could. But whether his own eyewitness testimony would have changed anything about the subsequent revulsion in London when the story hit the papers, we cannot know. A nation that had waited too long for news was now eager for someone to blame for the expedition's failure. That blame would not be dropped on Franklin. He and his men were lauded as tragic heroes. Instead, it would be placed on Rae and his Inuit witnesses.

Chief among the thunderers was no less a literary luminary than Charles Dickens. Alongside all those novels people are still forced to study at school, Dickens edited a weekly magazine called *Household Words*.* He used it to revile the Inuit ad hominem and to rage against their story. Describing the men of the expedition as "the flower of the trained English navy," he wrote, "the noble conduct and

---

*   Quite the compelling title . . .

example of such men, of their own great leader himself . . . outweighs by weight of the whole universe the chatter of a gross handful of uncivilized people, with a domesticity of blood and blubber."

Increasingly, the Inuit themselves would be held responsible for any cannibalism of the Franklin crew, despite John Rae's own response in Dickens's magazine. And despite cannibalism, or more correctly, anthropophagy,* being a long understood, if seldom talked of, custom of the sea.† Dickens stood at the vanguard of the opinion forming. "No man can, with any show of reason, undertake to affirm this the sad remnant of Franklin's gallant band were not set upon and slain by the Esquimeaux [*sic*] themselves," he wrote in the December 1854 issue of *Household Words*. "It is impossible to form an estimate of the character of any race of savages, from the deferential behavior to the white man while he is strong. The mistake has been made again and again; and the moment the white man has appeared

---

* Whereby the eater does not actively kill the eaten party, but survives instead on their dead remains.

† Famous examples predating Franklin's expedition include the sinking of the whaler *Essex* in 1820, which inspired Melville's *Moby Dick*, and powerfully dramatized in Ron Howard's movie *In the Heart of the Sea*. Of the seven men eaten, only one, Owen Coffin, was killed for food after losing a lottery among the survivors. Captain Pollard offered to protect him from his fate, but Coffin refused, allegedly saying it was his right to die so the others might live.

in the new aspect of being weaker than the savage, the savage has changed and sprung upon him."

According to this cheap rhetoric, Rae had to be wrong.

Lady Jane Franklin, who had done so much to lobby both for her husband's command of the expedition and then for its many rescue missions, simply refused to believe she had been widowed, and tried to prevent Rae's being paid the £10,000* allocated for the first person to recover evidence of the mission's fate. Though he finally got the money, he never received any recognition for his achievement. Unlike most of his contemporaries, there was no knighthood for the man who could survive like the Inuit.

Supported by Dickens, Lady Jane raised enough money from the public to fund yet another rescue mission† to be led by Francis Leopold McClintock, who had been a part of the aborted 1848 expedition with James Clark Ross. This mission, though it left England in 1857, would not begin its search properly until 1859.

It was McClintock's men who would find the Victory Point Note, mentioned previously, hidden in a small cylinder beneath a cairn. They would go on to find more artifacts strewn widely across King William Island: another body, whom they identified as the *Terror*'s

---

* Which would be worth approximately $1.5 million today.

† She had already spent £35,000 of her own money on rescue missions by this point, almost $6 million today.

gunroom steward, Thomas Armitage; a piece of paper which, if it had had writing upon it, was now bleached out; and, most intriguingly of all, a small pocket book belonging to Henry Peglar, captain of the foretop on HMS *Terror*. Now known as the Peglar Papers, these surviving documents have been pored over by Franklinologists for years. While most of them remain illegible, they offer tiny glimpses into the Expedition's suffering, and perhaps even a reference to the funeral of Franklin himself. Peglar writes: "The Dyer was and whare Traffalegar" [*sic*]—a misspelling of Trafalgar, a battle of which he and Franklin were the Expedition's only veterans.

In the years thereafter, mission after mission and theory after theory have been presented to try to explain what happened. Was the expedition brought down by scurvy, by lead poisoning resulting from the poor canning of their food supplies, by sheer bad luck? No one can say for sure.* The only reliable information we have comes from two sources: the surviving note, and the testimony of the Inuit. Despite the casual and dismissive racism of Charles Dickens, it is the latter that has brought us closer to the truth.

More Indigenous testimony was gathered over the next twenty years, this time by American expeditions

---

* It is now generally agreed that the weather they endured was unusually bad—the Inuit called the period between 1845–48 "the years without summers."

led by Charles Francis Hall and Lieutenant Frederick Schwatka. Both heard stories of the Inuit seeing and boarding the icebound ships. A man named Ikinnelikpatolek told Schwatka that he had first seen white men when he was a boy when he met ten strangers while he was fishing on the Back River. The next white man he saw was dead, lying in a bunk on a "great ship that was frozen in the ice." Hall, too, was told of a group of Inuit who came across a deserted ship at a place they called Utjulik—the place of seals. On board they found the body of a large white man lying on the floor, and described his remains as being "quite perfect." They said the place smelt very bad. The ships, they said, sank not long after they were abandoned, one of them in water so shallow that its masts remained visible for some time.

McClintock, too, gathered eyewitness accounts from the Inuit. His interpreter, Carl Petersen, was told by a woman that she had seen one of the wrecks in the winter of 1857–58, perhaps referring to the exposed masts that Hall was told of. She also said, "Many of the white men dropped by the way as they went to the Great River," which accorded with the bodies McClintock's people found. "They fell down and died as they walked along," he wrote.

And that was broadly that, and would have remained so were it not for the fact that, like many mysteries in human history, people have remained endlessly fascinated by the fate of the Franklin Expedition. What happened to them?* Why did it all go so wrong?† And what became of their ships?

The last question is the only one that can be answered with any certainty. That we can do so is largely due to the work of one man, an Inuit named Louie Kamookak.

Kamookak first learned of the Franklin Expedition in tales told by his great-grandmother and, over the course of thirty years, would go on to record and catalog all the Inuit stories about it. His goal was not only to find the wrecks of *Erebus* and *Terror*, but to find Franklin's grave and send him home. He would describe this work as akin to being a detective, realizing that, in their western tellings, many

---

* Ultimately, it seems they were undone by disease and starvation. In the early 1980s, Professor Owen Beattie of the University of Alberta launched expeditions to retrace the McClintock and other expeditions. One of his goals included examining the theory put forward by Sir Clements Markham that the expedition had been poisoned by the tinned food in their stories. Beattie exhumed the bodies buried on Beechey Island and, in addition to finding traces of scurvy, discovered unexpected quantities of lead in their bones. The bodies so far discovered also reveal pneumonia and tuberculosis, which were no doubt exacerbated by the scurvy. Later work on the bodies' tissue samples suggests their lead levels may not have been out of the ordinary at the time, and raise questions of a lack of zinc in their diet. In the body of Hartnell in particular, it would have been sufficient to suppress his immune system. But for all this scientific brilliance, we cannot escape the fact that they simply starved and froze to death, caught out in an environment they were ill-equipped to survive.

† It seems that Franklin chose to reach the Northwest Passage down the wrong side of King William Island. The route along the west of the island does not always clear of ice, while the east side route, taken by Roald Amundsen on his successful transit of the Passage, clears regularly.

of the Inuit's stories were riddled with so much mistranslation that they were next to useless. Like the Iñupiaq Harry Brower, Kamookak's work would come to bridge the gap between the traditional knowledge of the northern peoples and western academia.

Of course, when the young Louie was first captivated by the Franklin stories his grandmother told, he never heard the names of Franklin or Crozier or of any of the others. The Indigenous stories did not know them. They only told of white men, desperate, starving, and dying in the ice. It was only when he went to school that the white names of these men became known to him.

I worry about using the phrase "went to school" just there. It conjures images of an ordinary school bus, or riding your bike with a bookbag on your back. An image of a parent walking or driving you to the school gate, making sure you have your lunch and then kissing your cheek to embarrass you in front of your friends.

For Inuits of Louie Kamookak's generation, going to school holds no such cozy images. He grew up in an era of forced modernization. The Canadian government of the day called it a "civilising mission": taking children away from their families by force and placing them into schools far from their homes and, indeed, everything they knew, into supposed safe spaces where, denied the use of their own languages, many endured all manner of psychological, physical, and sexual abuse.

Just like Harry Brower, in faraway Alaska, Louie's parents must have feared for him, being pulled so far from his culture, from his language, from his people's stories. In an oral culture, these things matter deeply.* But Louie found a connection between these two worlds through Franklin.

Franklin remains a figure of huge stature in Canadian history. In the course of his several expeditions into the High Arctic, he and his men added more coastline to Canada's maps than any other man apart from George Vancouver. Though he was learning Franklin's name for the first time, in Franklin's story, Louie found an immediate link to his great-grandmother's stories and to his home.

But despite Franklin's stature as a Canadian hero, Kamookak found his growing fascination with the story unsated by the limited information in his school text books. Moreover, his education came to an abrupt end when he was just fifteen.† If he wanted to know more, his people's oral stories were the only place to go. And they would fire within him a curiosity for the expedition's fate that would last the rest of his life.

---

* I'd suggest they do in literate cultures, too, even though we've been raised not necessarily to notice it. But, when a grandparent dies, so many stories vanish from a family because no one ever wrote them down. Though this is purely anecdotal, when both my grandmothers died, so many stories vanished with them because we always relied on their presence to tell them. We did not have the practice of the oral tradition to remember them.

† At the time, the Canadian government briefly opted to stop residential schools for Indigenous children, so Louie found himself school-less at the beginning of the tenth grade.

Over the course of some thirty years, he would go on to catalogue oral Inuit testimony of the Franklin Expedition and its survivors. What sets him apart from those who tried this before is not just the sheer bulk of evidence amassed, it's that he was the first of the Inuit to do it. Unlike his predecessors, he was able to unravel the many inconsistencies of translation that lay within earlier work. More importantly, he had the local knowledge to understand the context in which these stories were told.

Within these tales, one frequently comes across the name Aglooka. In the stories gathered by John Rae, this is generally thought to refer to Francis Crozier, captain of the *Terror*. The name refers to the long strides taken by a tall (white) man. But in other stories, it can apply to James Clark Ross, who had explored this part of northeastern Canada several years before. Kamookak's particular knowledge and insight allowed him to give context not just to the story being told, but to the storyteller.

He realized that, if one knew the family group to which the teller of the tale belonged, one also knew where they ranged and hunted and, more importantly, when. All these details are contained within the Inuit pattern of storytelling not known to the likes of John Rae and his successors. This understanding of context ensured that Kamookak could add another level of meaning to the tales he gathered and, with it, more historicity.

Kamookak came from an Inuit group known as the Netsilingmiut—the people of the place where there is seal. Made up of interconnected family groups, they were nomadic hunters who ranged across more than 100,000 square kilometers of northern Canada to the west of Hudson Bay. It was they who made first contact with the white explorers searching for the Northwest Passage.

In addition to gathering their stories of those contacts, Kamookak began to unravel the family trees and inter-group connections that would furnish their stories with information beyond the tales themselves: location, based on where that group had ranged in a given season, and when. And therefore whom each Inuit band might have met in each encounter, be it men of Franklin's party, John Rae, James Clark Ross, and so on.

Such context is vital. There is something inherently romantic about oral evidence in history. The very idea of facts handed down from memory to memory over generations fires something soulful within, and we want it to be true. We yearn to believe in a historical Arthur or Beowulf or Achilles. But, with their oft-told tales, the context Louie Kamookak was able to provide is all too often absent. And this lack of an extra level of local knowledge bedevils those who would wish to use traditional knowledge as a valuable source.

It was this sense of romance that drove the archae-ologist Heinrich Schliemann to northwestern Turkey to drive a massive trench into a mound called Hisarlik to try to prove the existence of Troy, inadvertently destroying inordinate quantities of evidence from the late Bronze Age* in pursuit of something which he felt matched Homer's descriptions of the city. It drove him, too, to Mycenae where, having excavated several graves within the site's massive Cyclopean walls and finding royal bodies adorned with golden face masks, he claimed to have said, "I have gazed upon the face of Agamemnon."

But, all romance aside, we should note that Mycenae, Tiryns, and (arguably) Troy itself were all where the *Iliad* said they should be, not to mention the myriad sites across Greece that, though they never amounted to much in the Classical period, were named as significant in the *Iliad*'s famous Catalogue of Ships, considered to be the poem's oldest part. We should note, too, evidence of massive hide shields from earlier in the Bronze Age that would, if slung upon one's back, tap one on the back of the neck and ankles just as Ajax's was said to do. And we should note the famous find of the Dendra Armour, its bronze plates encasing all of the body but the calves and ankles when we think of the story of Achilles's heel.

We know from the archaeological layers of Hisarlik, what we call Troy 6 and 7a, that something monumental happened here. We lack the context to go further. But the work of Brower and Kamookak with scientists and

---

* Which might have helped us better understand this site at the apparent time of the mythical Trojan War.

historians, respectively, demonstrate the oral stories they both learnt and collected are stories of value still.

We should not dismiss them.

In his attacks on the Inuit testimony gathered by John Rae, Charles Dickens did exactly that. They were, to his mind, barbarous. Implicit in his words is the notion that, because they neither read nor wrote, their stories were not to be trusted. He missed what the Sumerians understood: that every great advance we make comes with a price.

Dickens could not fathom the notion that the price of literacy is memory. Literacy transforms memory's very nature. With it, we no longer need to hold in our heads the ideas we commit to books. Without it, we must depend on memory, both individual and collective, for everything we feel we need to know. Of course, we can learn these things by rote as an actor remembers their lines or a student studies for their exams, but we no longer need to. With the aide memoire of the library or the internet, we can simply look up that errant fact rather than committing it to memory.*

Nor could Dickens conceive that, in the High Arctic, the accurate transmission and memory of oral tradition is a matter of life and death. These stories contain not just

---

* You know this is true. How many phone numbers do you remember? Before you had a cell phone, you probably knew at least thirty by heart. For my part, I now remember just three: my wife's mobile, my late mother's house, and, for some reason, my late grandmother's place in Bournemouth in southern England.

a mythology to understand the world, but vital information about how to find one's way in a featureless white landscape, how to navigate by the stars, and where to find food in times of scarcity, information both Harry Brower and Louie Kamookak feared would be lost to Indigenous children who no longer spoke their own language nor desired to learn the old ways, as they had.

"When oral history is considered, a whole new Arctic reveals itself. The words often used to describe the space—mysterious, inhospitable, unnavigable—no longer apply. Instead, it becomes a place inhabited for roughly 4,000 years, with routes and trails well-traveled long before white men ever left their footprints in the snow."[*]

It is hardly surprising, then, that when Canada's Prime Minister, Stephen Harper, was able to announce in September 2014 that the wreck of the *Erebus* had been found, it was in no small part down to Kamookak's research.

It was almost exactly where the Inuit said it should be.

All of a sudden, he became something of a research celebrity, deluged with requests for more information. Always a man happy to share his knowledge, he had to acknowledge that, for those who wanted to learn from him, he wasn't hard to find. "I am the only Louie Kamookak," he noted.[†]

---

[*]   https://www.thecanadianencyclopedia.ca/en/article/arctic-exploration-editorial

[†]   Indeed. As his obituary four years later in the *Nunatsiaq News* said: "His like will not be seen again."

Kamookak's work would come to the fore again in 2016 when the wreck of the *Terror* was also found, again where the Inuit had orally recorded it. It would prove to be the last major Franklin find of his lifetime.

Like Brower, Kamookak bridged two worlds, that of the nomadic Inuit society of his birth and of the literate, settled, imposed culture of the West. His ability to embody that bridge led him to be called by the president of the Royal Canadian Geographic Society "the last great Franklin scholar."

Even as little as two months before his death, he was planning a new expedition to the north of King William Island with the goal of finding the so-called Supunger's Vault, as described in the testimony by an Inuit of that name to Charles Francis Hall in 1866.

Supunger told Hall that he and his uncle had found a marker stick and flat stones that had been carefully placed at a site not far from Victory Point. With some effort, they managed to move the stones to enter the vault, where they discovered a clasp knife and a partial human skeleton.

Kamookak hoped this story might contain the key to finding Franklin's grave, something that would have been the true culmination of his work: to return Franklin to his homeland and set his soul at rest. But, at the time of this writing, Franklin's grave remains undiscovered.

Of the 134 souls who put out from Greenhithe in May 1845, just five returned, discharged from the expedition at the Whale Fish Islands due to sickness.

Of the other 129, lost to the Arctic over the course of the next four years, just five have been positively identified. The wrecks have so far told us little more about their fate.

It seems that the *Terror* was abandoned when she sank, battened up in the hope that she may be recoverable. She drifted gently to the bottom to land on the seabed on her keel, leading one marine archaeologist to observe that if you raised her and pumped her out, she would probably float.

The *Erebus* appears to have been remanned and sailed further to the south, perhaps to get closer to the mouth of the Back River, until she could go further. And there, too, she was abandoned, leaving only tantalizing hints of what happened to her crew.

All that remains of their story is told by the Inuit, a testament to a mission that was never trained to survive on land, and to men who, for whatever reason—a lack of language skills, a belief in their innate superior ability, a blind hope they would never have to make it out on foot—never asked the local experts how to do so.

We will never know the circumstances that drove them from their ships, their sanctuary. We can only glimpse the intensity of their suffering in those final months on the featureless ice. But, while men of future British polar expeditions would die on the ice as well, they were never so poorly equipped to survive it again.

2

The Luxury of Cool

New York City—day. It's about 34 °C (93°F). The sun is blazing. It is sweltering even in the shade of the north–south Avenues' concrete canyons. Yet in just a few steps, we can walk into a climate-controlled restaurant, order a glass of iced water, or something stronger, and cool down.

We can do this almost all over the world; although, in some of the more far-flung places, climate control might be out of the question. But the cold drink from the fridge is almost ubiquitous.

That cold, cold bottle, frosted with condensation, in a hot, hot climate, is unremarkable. So unremarkable, in fact, that it is extraordinary. No other creature that we know of has ever had the ability to control the temperature of their food or beverages in the way we humans do. It is an everyday miracle we never even think about.

We should.

Our use of fire changed us. One can argue that it is one of our defining features. Our ability to cook has allowed us to extract more nutrition from our food, which changed the evolution of our species, driving the development of our brains. Which begs two questions: have we even thought about how our recent harnessing of the cold might change us? And why does it seem so normal? Perhaps, in the case of the second of these questions, because some of us have been using ice for millennia.

I began this book with a brief tale from Sumerian mythology. And it is to Sumer we must now return. For, as hard as it may be to conceive, standing on the walls of ancient Uruk and looking out over the barren, wind-swept, battle-scarred landscape beyond, the people of this first civilization were, as far as we know, the pioneers of humanity's use of ice.

The first recorded icehouse dates to Year 13 of the rule of Shulgi, King of all Sumer and Akkad, builder of the Great Ziggurat of Ur, a date which would, by our reckoning, fall somewhere around 2081 B.C.E. We know the icehouse was a big deal because the Sumerians like to name each year after something significant that happened within it. Year 13 was the year of the icehouse. It is described in the surviving cuneiform tablets as being twice as long as it was deep, and insulated with branches of tamarisk.

What we cannot know is whether this was Shulgi's innovation, or whether ice pits had been built previously, perhaps even before the founding of the Sumerian civilization. But if it was his or his engineer's idea, it would take more than 4,000 years for the use of ice to become as unremarkable as it is today. Yet historical sources for it remain limited perhaps because, even then, it seemed ordinary.

This is a fundamental obstacle with history: it is fickle. That thing we might be interested in might not have been of interest to the ancient writer. In the case of Sumer, the clay tablet we might dream to tell us what we hope to know could simply have been ground into dust by time. Or, in the case of more Classical texts, the papyrus scroll of our needs may simply not appeal to its copyist. So much knowledge has been lost. So many books by famous and important ancient writers we know of simply do not survive. Take, for example, our sources for no lesser figure than Alexander the Great: we know that contemporary sources wrote about him, but their books never made it through the centuries. While these works were cited by historians like Arrian and Quintus Curtius Rufus, they wrote at least 400 years after the events they described. This complicates things for those of us today who might wish to discern what really happened so very long ago. Looking into this distant past we, like Louie Kamookak, are detectives, but with even less to go on.

Nevertheless, we can still perhaps trace a through line of ice storage and usage from these distant times at least as far as the Classical era. References to icehouses dry up after Shulgi's momentous achievement, returning to the tablets some two hundred* years later, when icehouses are recorded in Mari, a Bronze Age kingdom in eastern Syria.

The question we cannot answer is whether there might have been an exchange of ice technology between the Ur of Shulgi and the Mari of Zimri-Lim, whom we meet next. Given the volume of correspondence that survives from the Near Eastern Bronze Age, we must consider it a possibility, not least because of the this tantalising fact: Shulgi was married to a woman named Taram-Uram, the daughter of Apil-Kin, Shulgi's contemporary and ruler of Mari.

Mari grew up on a trade crossroads between Mesopotamia and Babylon. Not only is it considered to be one of the earliest planned cities, it was also a center of technical innovation. Its people were great canal builders, both for irrigation† and navigation, constructing one that ran 126 km past the city, allowing traders a more direct route that bypassed the long and winding Euphrates. The icehouses crop up towards the end of Mari's time of influence, the late nineteenth century B.C.E., when the city was ruled by the so-called Lim Dynasty.

History does not seem to have run smoothly for the Lims. Their first ruler, Yaggid-Lim, successfully passed the throne to his son, Yakhud-Lim.‡ Despite various

---

* Or so.

† They built a good sixteen km of canals for this.

‡ This despite Yakhud-Lim's having to act as a hostage to the king of Ekallatum in the years leading up to his father's death.

successes, including expanding his city-state's irrigation and refortifying his wall, Yakhud-Lim found himself becoming vassal to the more powerful city of Aleppo, then of the Mesopotamian city Eshunna,* and then, having reasserted his independence one more, at war with the Assyrians. All this before being assassinated by his son.

That would have been that for the Lim Dynasty were it not for Zimri-Lim, Yaggid-Lim's apparent grandson, in approximately 1776 B.C.E. And it is to Zimri-Lim that we must turn to find our ice.

Having cemented his relationship with Aleppo† by marrying its king's daughter, Zimri-Lim sought to restore Mari to its former glory. He did so through diplomacy, marrying his daughters to neighboring kings, allying with the powerful kingdom of Babylon to his south, and, thanks to the frequent letters exchanged with his married-off daughters, keeping himself abreast of everything around him through what is best described as a familial intelligence network. His vast archive of correspondence and official records was uncovered by archaeologists in the 1930s. Written in Akkadian, the official language of Near Eastern diplomacy, these tablets—more than 22,000 of them—shine an extraordinary light into the life and workings of the Syrian mid-Bronze Age. And it is one of these tablets, the so-called

---

* Which lies about 32 km northeast of Baghdad.

† At this time, Aleppo, sometimes rendered Halab, was in effect the capital of a kingdom called Yamkhad which lay to Mari's west.

Tablet of Zimri-Lim, now housed in the Louvre, that tells
of the foundation of an icehouse in Terqa, not far from Deir
ez-Zor on the banks of the Euphrates.

It seems that the Terqa icehouse was not unique.
Zimri-Lim claims to have built similar icehouses across
his kingdom. But, while we know from the tablets that they
existed, we have no idea what they looked like, how they were
stocked with ice, nor how they were maintained. Our best
guess suggests that they might have been similar to later
examples built by the Persians. This would seem to make
sense. Though the surviving ancient Persian icehouses,
called *yakhchals*, were built around 400 B.C.E., much later
than those of Zimri-Lim, there is no reason that their
technology should not have existed a good 1,300 years later.

A *yakhchal* works by evaporative cooling. In the dry
desert air, the temperature drops fast after sunset, often
going below freezing at higher altitudes. The *yakhchal's*
domed structure allows this cold air to pour inside and into
the pit within, while the conical walls draw warmer air up
and out. In addition, the structure is built of a particular
mortar called *sarooj*, made to a specific recipe that includes
sand, goat hair, clay, egg whites, ash, and lime. It is both
water resistant and an excellent insulator.

A *yakhchal* gives you ice and cold food storage under
specific regional conditions—were there higher humidity
here, it simply wouldn't work—and those conditions exist
on the site of Mari. But, though it seems highly likely, we
cannot know if the two cultures used a similar or the same

technical idea to make their ice. There is a missing link between them in the chain of ideas, so we cannot connect them. We can only suppose.

Nor can we connect the Persian use of ice with that of Ancient Greece. We know, for example, that snow was available in the markets of late fifth century B.C.E. Athens. There is a story recounted by the rhetorician Athenaeus about the comedian Diphilus, whose many plays only survive in fragments, in which he attends a dinner at the house of a woman called Gnathaena where the wine is cooled with snow sent by one of her lovers. We know, too, from Chares of Mitylene,* that when Alexander came to Petra after his conquest of Nabataea, he ordered the construction of ice pits, to be filled with snow and protected with oak branches. Lacking the climactic conditions that allowed the Persians to build their *yakhchals*, the Greeks took a different tack to storing the cold. We find it described in Plutarch, though the pits he records were covered with straw and cloth.

Similar techniques were practiced in Italy from the Roman period into the nineteenth century. In his book *The History of Ancient and Modern Wines*, Alexander Harrison quotes a Mr. Lumsden, who describes the collection of snow for use in Rome at a site known as Hannibal's Camp:

---

* As recorded in Athenaeus's work, *The Deipnosophistae*, for Chares's *Records of Alexander*, a contemporary account of Alexander's life, does not survive.

On this dry plain they dig pits, without any
building, about fifty feet deep, and twenty-five
broad at the top, in the form of a sugar loaf, or
cone. The larger the pit, the snow, no doubt,
will preserve better. About three feet from the
bottom they commonly fix a wooden grate,
which serves for a drain, if any of the snow
should happen to melt, which otherwise would
stagnate, and hasten the dissolution of the rest.
The pit thus formed, and lined with pruning
of trees and straw, is filled with snow, which is
beat down as hard as possible, till it becomes a
solid body. It is afterwards covered with more
prunings of trees, and a roof is raised in form
of a low cone, well thatched over with straw.

So, not wildly dissimilar to Alexander's pits at Petra.
Ice so stored was then sold in shops or hawked on the
streets of Ancient Rome, much to the disapproval of Seneca
who, in his work *Naturales Quaestiones*, lamented its use as a
"true fever of the most malignant kind." Sniffy old Seneca
aside, ice was a serious luxury business, even if its transportation from store to city often caused it to be spoiled with dirt,
which led to the invention, described by Pliny the Elder, of a
vessel in which previously boiled water could be surrounded
with this snow, without contamination, and left to freeze.

Such references to ice remain rare in the ancient canon.
But this should not make us think that the thing they

describe was rare as well, even if it was not available to everyone. We have to tease the quotidian from our ancient sources. The everyday is most often mentioned as an aside, for it is something the intended reader already knows and doesn't need explaining to them. As L. P. Hartley wrote,* the past is a foreign country, and we—the unintended readers—are frequently left to deduce from scant evidence just how differently they did things there.

But, despite the paucity of sources, we can find ice used almost everywhere from the Song Dynasty of China at the turn of the first millennium to the ice pits of Allahabad in northern India. Indeed, legend has it that Kublai Khan, founder of the Chinese Yuan Dynasty, introduced the traveler Marco Polo to ice cream—a personal favorite of the emperor, who decreed that no one outside his family was allowed to make it.†

The ancient Chinese seem to have harvested and stored natural ice in a similar manner to that which we'll dig into in the next chapter. The people of Allahabad, lacking ice to harvest, had to make it evaporatively,‡ pouring boiled water into specially dug pits at sunset, and then gathering the frozen bounty for storing before the following sunrise.

---

*   In his book *The Go-Between*.

†   There is a second legend attached to this story which claims that Marco Polo then brought the secret of ice cream back to Italy.

‡   As recorded by Sir Robert Barker in 1775.

Meanwhile, around the Mediterranean, those in search of something cooling relied upon gathered and stored snow, as the Romans had before them. We have accounts of the Mamluk Caliphs and Sultans, who ruled Egypt and much of the Levant from 1250 until 1517, importing snow from the Lebanon to Cairo in the thirteenth and fourteenth centuries. In the same period, we have accounts of snow pits—presumably similar to those made in Italy—being built on the orders of King Charles III of Navarre. In France, too: at Vauclair Abbey, south of Laon in northern France, an icehouse has been excavated that dates from this period. And in the Near East, Saladin, Sultan of Egypt, is not only said to have sent ice to King Richard of England when the latter was struck down by a fever* but, after his decisive victory over the Christians at the Battle of Hattin in 1187, he famously executed the crusader Raymond de Châtillon after Raymond had drunk from a cup of iced rose water that had not been given to him.†

Though by no means as ubiquitous as it would become, ice and ice use were very much present in the Medieval world, a desired luxury in sophisticated circles. But it was still something remarkable. In 1494, the Milanese canon Pietro Casola undertook a pilgrimage to Jerusalem, which he recorded in a journal detailing the places and

---

*   Most likely a myth, by the way.

†   The scene is dramatized in Ridley Scott's film *Kingdom of Heaven*, and was recorded by the Persian scholar Imad ad-Din al-Isfahani, who was present at the time.

churches he visited. Writing of his arrival in Jaffa, he tells of gifts brought by a group of locals to his ship's captain. Included in the gifts was a sack of snow. "It was a great marvel to all the company to be in Syria in July and see a sack of snow," he wrote. "It was also a comfort to many, because some of the snow was put into water—which was hot—and cooled it." It would take another two hundred years for ice use, at least in France and the Italian states, to really take off. And it was the people of Florence, then ruled by the Medicis, who were at the forefront of it, in no small part thanks to the architect and stage designer Bernardo Buontalenti.*

Buontalenti was a genius. In addition to architecture, he could turn his hand to hydraulic engineering, weaponry, the making of fireworks, mathematics . . . he even painted a bit, too. While he is erroneously considered by some to have invented gelato, he was probably the first to bring ice to the masses. As well as building private icehouses, including the one in the Boboli Gardens, Buontalenti built several around the walls of the city to contain snow and ice for sale to the general public, having first obtained the concession to sell it from Grand Duke Ferdinand I himself in 1598. He would hold the snow monopoly until his death.

It seems slightly surprising that it took so long for a public ice venture to take off, for the Florentines enjoyed

---

*   c. 1531–1608

chilling their wines at least as far back as the 1340s. *La Cronica Domestica di Messer Donato Velluti*, written between 1367 and 1370, describes barrels of Tuscan white wine in ice. By the time of Grand Duke Cosimo I,[*] we find accounts of huge silver wine coolers weighing almost 12 kg, all this despite most doctors of the time claiming that drinking wine chilled with snow was actively harmful. Pushback came from Spain, where a doctor called Nicolas Monardes wrote a short work entitled *Libro de la Nieve*, published in Seville in 1571. More than anything else, it is a plea for a regular supply of usable snow to the city.

Writing about it in her book *Harvest of the Cold Months*, Elizabeth David wonders whether its translation into Italian in 1589 might have happened with the sponsorship of Cosimo's son and heir, Prince Francesco. Not only did the book rebuff the prevailing medical thinking of the day—Monardes is adamant that ice is good for you—but snow was becoming a source of revenue. If Buontalenti had the ice concession for Florence, what about the other towns and cities under Medici control? Surely their ice concessions could be sold, too, and the trade could be taxed.

Monardes, however, was not to have the last word, and the Italian doctor Piero Nati published his own tome, proclaiming that icing wine was bad, warm wine unpleasant, and proposing a middle course between the two. The back and forth would go on in European

---

[*]   Who ruled from 1537 to 1569.

medical publishing for decades, but for all the difference it made to the public taste for cold drinks in Italy, France, or Spain, these books amount to little more than a curious waste of ink and paper. Especially now that Francesco had acceded to the Grand Ducal throne, developed an interest in Turkish chilled sherbets, and wrote to the Venetian Mafeo Veniero to see if he could find him the recipe.

Elizabeth David digs deep into this,* hunting for the origins of Tuscan *sorbette* and *gelato*, so wrapped they are in myth and presumption. Even she was unable to discover what Francesco might have done with these recipes or, indeed, if he ever received them. But the use of snow as a cooling agent was very much the hot thing of the day.

Snow has its flaws. Even when compacted into as solid a form as possible, it easily gathers dirt and impurities which, when added to a drink, will be imbibed. Ice is so much purer. Moreover, it takes a lot of work to prepare snow for transportation. It has to be packed down and compressed. It melts more readily. So it is to the making of ice rather than shifting snow that Buontalenti turned. In 1603, we find new words in the documents of the Medici's office of works, words like *pescheria del diaccio*: man-made ice reservoirs built in the Appenines to supply new ice wells of Buontalenti's creation.

---

* In *Harvest of the Cold Months*, page 11, in which she tells of a letter written by Grand Duke Francesco de' Medici to Mafeo Veniero in Venice, asking for "the *ricetta delle sorbette*"—the recipe for making Turkish style sherbets.

By the time of Buontalenti's death in 1608, the Florentine ice trade was big business. Not that he saw much of the benefits. Not long before he died, he suffered a fall at work from which he never properly recovered and, at the beginning of 1606, found himself forced to petition the Grand Duke for a pension just to put food on the table—a sad and impoverished end for a man who had given Florence so much. It was his work that began the shift in ice's status from an item of high luxury to a widely available necessity.

Just sixty years later, in the poem *Bacco in Toscana*, Francesco Redi would describe snow as "the fifth element," demanding in verse that ice be brought to him from the Boboli Gardens store and crushed to the finest powder to cool his drink, "For I cannot swallow warm wine." He deploys the phrase "*la quinta essential*," from which we derive the word "quintessential," precisely to demonstrate ice's importance. It had become *necessarissimo*, extremely necessary.

The French, who embraced the use of ice later than their Italian and Spanish neighbors, would do things a little differently, drawing influence, according to the fabulously named eighteenth century social historian Pierre-Jean Baptiste Le Grand d'Aussy, from Turkey. Icehouses were a big deal in Ottoman Turkey, the concept of them almost certainly arriving during the earlier Seljuk Empire, which was essentially Turko-Persian and lasted from 1037–1194. In Kayseri, for example, in Central Anatolia, ice would be harvested in the winter from the slopes of Mount Erciyes, and be stored in pits or cellars for use in cooling drinks and food preservation in what may be considered to be the very first attempts at making iceboxes.

Seljuk and Ottoman Turks were also employing ice medically, not only to cool feverish patients but to store medical preparations. Several hospital sites from across the country are known to have had their own ice pits. And the good Dr. Monardes tells us that, in Constantinople, one could buy snow year-round.[*] So, when the French naturalist Pierre Belon set out from Greece across Asia Minor in 1546 on a five-year journey that would also take him to Egypt and east to Persia,[†] he was delighted to be served "cool sorbets,"[‡] or what we might now describe as fruit sherbets—the very same as those for which Grand Duke Francesco wanted the recipe. Belon describes the Turkish approach to icehouses in some detail, and as a good thing. But the French would not take up the use of ice until the end of the sixteenth century. Le Grand d'Aussy says that, prior to that, it was the French habit to serve their drinks warm year-round, which

---

[*]   In his 1569 book *Tratado de la Nieve y del Beber Frio*. He also states that, in addition to Constantinople, snow and ice were being sold "in all the German states, and in Flanders, Hungary and Bohemia." In the same passage, he also describes snow being transported from Flanders to Paris. If he is correct, and there is no reason to doubt him, it would explain where the ice Henri III is alleged to have introduced to the court of France was coming from.

[†]   Recorded in his book *Les Observations de Plusieurs Singulaitez et Choses Memorables en Grece, Asie* in 1553.

[‡]   As he described them.

Elizabeth David sees as a logical explanation for what she describes as "the violent hostility to the idea of iced wines and water for a long time expressed" by them.* This goes entirely against the popular myth that Catherine de' Medici brought ice cream and ices with her from Florence when she married Henri, then Duc d'Orleans, in 1533.

Like all good myths, this story has a certain charm and is still widely believed. But there are a couple of reasons why it is utterly bogus. For a start, she was just fourteen when she was sent to France to marry the second in line to the throne. And, under French law, it was required that she become a French subject and that her entire household become French. Perhaps we should imagine her, like Marie Antoinette on her journey from Austria to marry the Dauphin,† being stopped at the border and stripped of all things Tuscan before being dressed again as a lady of France.

Secondly, and probably more importantly, ice cream had yet to be invented, at least in Western Europe. For that, we must leap forward in time to 1674 to find the first recorded recipe in French, in a book by the chemist and cookery writer Nicholas Lemery.

You cannot make ice cream without understanding how to create an endothermic reaction to make the cream mixture cooler than the ice itself. This is a little bit of a

---

* Again, in *Harvest of the Cold Months*.

† As described in Antonia Fraser's biography of her and shockingly depicted in Sofia Coppola's movie.

challenge, since cream freezes approximately half a degree Celsius lower than water. Until you know how to make your ice colder than freezing, all you're going to end up with is cold cream. The secret to this everyday miracle does not seem to have reached Europe until the sixteenth century, even though it had been recorded elsewhere much earlier: salt.

Of course, human beings have been using salt for purposes other than making ice cream since before they walked out of Africa. But mixing it with ice to create a brine in which to then freeze something else in a container placed within it—well, that was comparatively new.*

It works because salted water freezes at a much lower temperature than fresh.† With a highly concentrated brine, you can achieve a freezing point as low as −21.2°C (−6.1°F), which explains why you need a hell of a lot of salt to cool your icy solution enough to make your ice cream. This offers up another possible reason why the science required to make it arrives well after people began messing about with ice's cooling properties in their drinks and desserts: Salt was not cheap. Salt monopolies and salt taxes persisted around the world in various forms well into the twentieth

---

*   The first reference to such cooling appears in the *Panchatantra*, a collection of Indian poems dating from the second century C.E.; it was also described by the Syrian physician Ibn Abi Usaibia in the thirteenth century.

†   A few facts on this: technically, a brine has a minimum of 50 parts per thousand of salt within the water. To give you an idea of how salty that is, sea water on average contains just 34.7 PPT.

century. Given its necessity for human nutrition, not to mention its value as a food preservative, perhaps it was felt that it would be a waste to use so much of it just to freeze ice cream.* It would be another two hundred years or so until different endothermic reactions, like dissolving ammonium nitrate in water, would be used in place of this basic version.

All of which is a long way of saying that, when we think of the sumptuous "ices" served in the courts of King Louis XIV of France and others,† they are closer to slush puppies than Ben and Jerry's. And they were still by no means available to all. But, little by little, the use of ice in desserts and drinks was spreading among the wealthy of the seventeenth century, if more by fashion than a Tuscan sense of *necessarissimo*. Coincidentally perhaps, this growing demand for ice took place at the same time as a phenomenon that ensured its plenty, the so-called Little Ice Age.

This was first identified by Dutch American geologist François Emile Matthes, who coined the term in 1939, and dated it from the sixteenth to the nineteenth centuries. As brilliant as Matthes undoubtedly was—an MIT graduate who taught at Harvard and went on to become the senior geologist at the US Geological Survey—this seems to be wrong. But, in Matthes's defense, even today there is no

---

* At least until they'd tasted the results . . .

† As detailed in Elizabeth David's *Harvest of the Cold Months*, chapter 5.

consensus on the Little Ice Age's timing or duration, nor indeed on its causes. This is something modern scientists are determined to pin down. Understanding the dynamics behind historic climate change offers insight into the challenges we face today. As the most recent example of such climate change, the Little Ice Age offers some of the best evidence for its natural* mechanisms.

A recent article from 2012[†] argues that the Little Ice Age lasted much longer than Matthes and others had previously thought. Using ice cores drawn from the Canadian Arctic, Greenland, and Iceland, they demonstrate, if not a direct correlation, then a coincidence of timing between an expansion of Arctic Ocean sea ice and spectacularly powerful volcanic eruptions on the other side of the world.

The article's authors wrote:

> Our precisely dated records demonstrate that the expansion of ice caps after Medieval times was initiated by an abrupt and persistent snow-line depression late in the 13th Century, and amplified in the mid-15th Century, coincident

---

*   As opposed to man-made.

†   Titled "Abrupt Onset of the Little Ice Age Triggered by Volcanism and Sustained by Sea-ice/Ocean Feedbacks," written by the team of Gifford H. Miller, Áslaug Geirsdóttir, Yafang Zhong, Darren J. Larsen, Bette L. Otto-Bliesner, Marika M. Holland, David A. Bailey, Kurt A. Refsnider, Scott J. Lehman, John R. Southon, Chance Anderson, Helgi Björnsson, and Thorvaldur Thordarson, and published in *Geophysical Research Letters*, vol. 39, L02708, 2012.

with episodes of repeated explosive volcanism centuries before the widely cited Maunder sunspot minimum (1645–1715 A.D. [Eddy, 1976]).*

Forget Edward Lorenz's so-called Butterfly Effect:† forty cubic kilometers of dust and debris suddenly flung into the atmosphere is going to have an impact.

The problem for the article's authors (there are thirteen of them, so please check the footnotes for their names) was that, while they could see within their ice cores that a sizable volcanic eruption had acted as a trigger for the Little Ice Age, they didn't have a culprit. It was only a year after they published that another group of scientists would unmask it.

The Samalas volcano in Lombok, Indonesia, exploded in 1257. Its effects were catastrophic. The temperature dropped; crops failed; people died—we can see it in the archaeology and surviving texts from the following year. Add to this a similarly sized eruption almost two hundred years later—this time in the South Pacific and of such a scale that it blew the islands of Epi and Tongoa in the

---

* The Maunder sunspot minimum being a seventy year period with very few observed sunspots. Within it, between 1672 and 1699, fewer than fifty were seen, despite a systematic series of solar observations made by the astronomer Giovanni Domenico Cassini at the Observatoire de Paris.

† So beautifully described by Terry Pratchett and Neil Gaiman in their book *Good Omens* thusly: "A butterfly flaps its wings in the Amazonian jungle, and subsequently a storm ravages half of Europe."

Republic of Vanuatu completely apart,* leaving a crater some twelve kilometers by six in diameter, and an extra kilometer deep for good measure—and you have the recipe for several centuries' sustained global cooling.

The force of an eruption is measured by the Volcanic Explosivity Index, devised by Chris Newhall of the United States Geological Survey and Stephen Self from the University of Hawaii back in 1982. It is a logarithmic system, much like the Richter scale used to quantify an earthquake, and it runs from zero to eight.† A seven blows between 10–100 cubic kilometers of debris into the atmosphere.

Samalas was a Level 7. So, too, was the one that created the Kuwae Caldera that divided Epi and Tongoa. And so was the eruption of Mount Tambora in 1815, which caused a so-called "year without a summer" towards the end of the Little Ice Age. Though we have experienced some seemingly gargantuan eruptions over the course of the last century or so, like Mount St. Helens—a Level 5—and Mount Pinatubo—a Level 6—we have been fortunate not to see a Level 7 since.

The debris spewed from a volcano has specific effects on the atmosphere, dependent on volume and the eruption's location. Should it take place at northerly or southerly latitudes, its effects will be more localized within its hemisphere. However,

---

* Previously, they had been one landmass.

† There have been no Level 8 eruptions since Mount Taupo in about 26,500 B.C.E.

the closer to the equator it takes place, the more likely its detritus is to be spread on atmospheric currents around the world. And once within those currents, two key factors contribute to a global cooling. The first is the presence of tiny dust particles of ash and dust that shade the earth from the sun's heat and light. The second is sulfur dioxide, which reacts with water vapor high in the atmosphere to create sulphuric acid droplets. Captured in the high stratosphere, these reflect solar radiation back into space. They can stay in the atmosphere for years, cooling the earth below before they become large enough to fall to the ground.

Of course, none of this was known to those who endured the summerless years that followed these eruptions. To them, their crop failures must have seemed like an act of a malevolent god or some kind of punishment for their sins. Whereas we, today, are able to study the interconnectedness between such seemingly discrete events.[*]

The upshot of all this[†] is that there was a lot of ice about. And yet not everyone was to embrace its use as quickly and eagerly as the Florentines of the royal court

---

[*]   One of my favourites is the linking of an orphan tsunami, which hit Japan on January 27, 1700, with Native American stories from the Pacific Northwest about a terrifying earthquake and tsunami that ravaged their lands. Thanks to work across the disciplines of Japanese history and global geoscience, we now know that this was the last time the Cascade Fault, which lies just off the coast of Oregon and Washington, went off. According to most analysis, it is now overdue for another quake. And the effects of it could be terrifying.

[†]   Not to be too flippant about it.

of France. Like the British. Even today, people—mainly
Americans—remark upon the British people's reticence in
the use of ice. Back in 1943, the legendary British filmmaker
Anthony Asquith* teamed up with the actor Burgess Meredith to make a GI training film entitled *Welcome to Britain.*†
In a telling sequence, Meredith introduces the viewing
trainees to the concept of a pub, explaining that it isn't
like an American saloon. As he's about to open the door to
one, he turns back to camera and says: "And incidentally,
the beer isn't cold in England. They don't like it cold, and
they don't have any ice. So, if you like beer, you better like
it warm." It's a reticence that survives to this day despite, as
the drinks writer Henry Jeffries records, a fad for drinking
claret chilled in the late nineteenth century. As Jeffries
goes on,‡ ice use never became ubiquitous in Britain. But,
of course, it was there, even if no one was bothering with it.

The first recorded purpose-built ice or snow store in
England dates from 1619, built for the future King Charles I,
then Prince of Wales, at Greenwich Palace, which had
been assigned to him by James I after the death of his

---

* Please excuse the personal aside here but, for my money, Asquith
is a massively underrated talent in the history of British film. His
pictures are meticulously made, and incredibly emotional. He
deserves to be placed in the pantheon of visionary British directors
alongside Alfred Hitchcock, Michael Powell, Nicholas Roeg, and
others. Seriously, if you take nothing else from this book, please
go and watch some of his films. He was a true master of the art.

† Also known as *How to Behave in Britain.*

‡ In an article he wrote for the *Guardian*, February 13, 2015.

mother, Anne of Denmark, earlier that year. Despite the prince's relatively early adoption of ice, it would take a while to catch on with both the aristocracy and the broader public. Indeed, it's not entirely clear how well the English understood the concept of refrigeration. In his architectural palimpsest on icehouses, Tim Buxbaum records Francis Bacon's surprise in 1626 that ice could keep poultry fresh, at least for a while.

This tale, recounted by the English natural historian and philosopher John Aubrey, is most likely apocryphal. Bacon had, in fact, been studying the effects of the cold for some time, and recommended in his posthumous book *Sylva Sylvarum* that the student of the natural world should have "a conservatory of snow or ice" on hand. Whether he had or not, we don't know, such (as we've seen) is the nature of hunting for ice in our sources.

It's worth dwelling on the gradual English adoption of ice, if only because from them the principles of icehouses would travel to America. But it's difficult: as with so many things to do with ice, information about them from the period remains thin on the ground. All too often, they're like ghost buildings, unremarkable and unrecognized, broadly forgotten after their replacement by later technologies. But we can best tie their introduction to the era following the Restoration.

At the forefront of the new fashion was, hardly surprisingly, the new King Charles II, who had icehouses built at his newly laid out Green Park and, in 1662, at Hampton Court. And where the king led, the peers of the realm would eagerly follow. At Blenheim Palace, the royal gardener Henry Wise, who created its original formal gardens from 1705 onwards,* sited and built two icehouses in the grounds in 1707. Wise's mentee and successor as royal gardener, Charles Bridgeman, would incorporate icehouses into his "wilderness garden," built in Richmond in 1734,† and into the gardens he designed for Lord Cobham at Stowe, which would be described by the poet Alexander Pope as "work to wonder at."

Bridgeman was a pioneer of a less-structured style of garden and landscape design, the so-called *jardin Anglais*. He laid out carefully created views over the countryside, geometrically shaped lakes, canals, avenues, and kitchen gardens. In his addition of an icehouse with a viewing mount at Richmond, he may have been imitating a similar construction from the gardens of the Governor's Palace in Williamsburg, Virginia.‡ Built at around the same time as Henry Wise was sitting his within his original garden designs

---

* Before they were ripped out and replaced with those we know today by Lancelot "Capability" Brown.

† On the site that would later become the famous Royal Botanic Gardens at Kew.

‡ This, from an implication in Buxbaum's short book *Icehouses*, page 29. It is very hard to track the exchange of ideas about ice usage and icehouse construction in a way that usefully demonstrates their evolution. Clearly, architects and garden designers were imitating ideas seen both abroad and within their own countries. But where we do find written accounts, they tend to be in written correspondence and often take the form of "I have heard that X did Y," as we will see with Frederic Tudor in the next chapter.

at Blenheim Palace, the Governor's Palace icehouse must be among the very first in what is now the United States.

As an integral part of fashionable garden design, these late seventeenth–eighteenth century private icehouses were not simply functional creations. They were a part of their designers' ornamentation of their curated landscaping, mainly hidden below ground with a decorative folly above, or hidden from view completely. But inside, their structure was broadly the same.

A typical icehouse of the period would feature an inverted cone chamber, shaped much like a sugar loaf, into which the ice or snow would be packed during the winter. They would have a drain at the bottom to ensure that melt water didn't spoil the house's frozen contents. The entrance would be laid out to try as far as possible to prevent warm air from creeping within. And finally, the ceiling would be insulated with straw or sawdust to keep in the cold.

More important than the design of the icehouse was the tight packing of its contents. "The closer it is packed, the better it keeps," wrote Robert Morris, the so-called financier of the American Revolution. His ice-house, at his estate in Philadelphia, so impressed George Washington that he asked Morris for detailed instructions on how to build and use one so that he could install his own at Mount Vernon some time around 1790. He was not the only founding father to partake: Thomas Jefferson would add an icehouse of his own design to his estate at

Monticello in 1806, and build another at the President's house in Philadelphia.

In building them, these aristocrats of the new United States were following a fashion promulgated by kings like Charles II and Louis XIV little more than a hundred years before. They built icehouses, and so did the noblemen of their courts. An icehouse, even two, became a signifier of wealth, success, and status. For both Louis and Charles, in their day, were exporters of taste, the influencers of a golden age.

Which raises a question: Why should their influence have more impact than the Medicis in Florence, the Chinese or Turkish emperors, or even their contemporary kings of Spain or Portugal? For if the driver of this influence is the growth of empire in the late seventeenth century,* and the Spanish and Portuguese were the pioneers of European empire building, why not them?

In the case of China, Charles and Louis's contemporary, the Kangxi Emperor of the recently established Qing Dynasty, faced a number of wars of consolidation and struggles for control of a fractious empire, all of which would allow him to lay the foundations of the so-called High Qing era of prosperity. For our purposes, ice to him and his court was not a novelty.

Similarly, for the Ottomans, ice was nothing new. The ever-changing rulers of the Near and Middle East had

---

*   As it almost certainly is . . .

been storing and trading ice for centuries. Fashion depends upon that feeling of must-have, up-to-the-minute now-ness, and the need to fit in.

As for the Medicis, for all their power in Tuscany and influence in Europe, theirs was not an imperial Duchy. Their ice would remain *necessario* . . . to them.

Meanwhile, as the Spanish and Portuguese* set about colonizing the tropics, we know they were aware that it was quite the thing to know how to store ice for the summer. But this is utterly useless information if you have nowhere to source the ice you might have planned to store. This broadly holds true for the French as well. Louisiana, named for Louis XIV, is not noted for its snowy winters, even though the Canadas are.

Charles, however, held New England, the Carolinas, and those other territories that would combine to form the thirteen colonies that would fight for their independence from George III. And it would be in New England— where ice could be both harvested and stored—that the shift in its status from a luxury to an everyday global com-modity would be imagined and achieved.

---

* If I can make a massive sweeping statement . . .

3

The Ice Man Cometh

S t. Gregory of Tours began his sixth century work, *The History of the Franks*, by writing: "A great many things have happened, some of them good, some of them bad." Which is possibly the pithiest summary of history ever put to paper,* descriptive of any year you can think of.

Take, for an example, 1862. The American Civil War was raging. Britain annexed Lagos Island in Nigeria. The French invaded Mexico. Finland and New Zealand opened their first railways. Victor Hugo published *Les Misérables*. The composer Claude Debussy was born, as was the artist Gustav Klimt, the writer Edith Wharton, and the "Elephant Man" Joseph Merrick. And, to much less fanfare, a barman by the name of Jerry Thomas published a pioneering book called *The Bar-Tenders' Guide*.

No one had ever written a guide to mixing drinks before.

While Thomas may be little known outside the profession, he remains revered as the godfather of bar-keeping.

---

* It may just be my favorite first line of any book ever . . .

His peers nicknamed him "The Professor," and every mixologist or barkeep working today owes the foundation of their knowledge to his work.

The Bar-Tenders' Guide essentially codifies what had been, up to its publication, an oral tradition of drinks recipes, which Thomas collected and recorded as he worked his way around the United States, from Connecticut to California. He has this to say on the matter of ice: "Ice must be washed clean before it is used."

And that's it.

This may not seem striking to those of us used to a crystal clear, icy cold martini or a perfectly mixed negroni with a single, large cube floating within, but just fifty years prior to Thomas's book's publication, ice remained a luxury item. To him, it had become so commonplace he barely mentions it.

In fact, ice had become so ordinary that, when you crawl through old cocktail books or more recent works on cocktail history, not one of the authors can tell you* when

---

* Or can be bothered to tell you. . . . Having said that, *Difford's Guide*—one of the key modern cocktail bibles—offers an excellent guide to the origin of the word "cocktail." The bottom line—nobody really knows. The first found usage of it dates to 1798, while the first definition appears in 1806, coincidentally, the same year that Tudor started his ice business. It appears in *The Balance and Columbian Repository*, and reads: "Cock-tail, then, is a stimulating liquor, composed of spirits of any kind, sugar, water, and bitters—it is vulgarly called a battered sling, and is supposed to be an excellent electioneering potion, in as much as it renders the heart flout and bold, at the same time as it fuddles the head. It is said also, to be of great use to a democratic candidate: because a person having swallowed a glass of it, is ready to swallow anything else." Ice is not mentioned.

ice first appeared in a mixed drink, even though almost every extant cocktail recipe requires it.[*]

The chilling of wine[†] notwithstanding, a shift in drinking culture had taken place—an American shift which would prove to have far-reaching effects that went far beyond the doors of the saloon. And the driving force behind it was a determined and diminutive Bostonian called Frederic Tudor.

Tudor's particular genius lay in a streak of stubbornness so deep that it would put a terrier to shame. But there was nothing about his early life to suggest he might change the world.

The Tudor family was a relatively new addition to New England. Frederic's grandfather, John, had emigrated from Devon, in the southwest of England, with his widowed mother in 1715. He was just six years old. John prospered in the New World, enough to be able to send his youngest son, William, to Harvard[‡] and to acquire both a Boston townhouse and a hundred-acre country estate called Rockwood in Saugus, Massachusetts.

Its grounds contained an icehouse.

William, too, made a success of himself. After his stint at Harvard, he went on to study further in the law offices

---

[*] The notable exceptions being those, like the whiskey toddy or an Irish coffee and their friends, which are meant to be served hot.

[†] Discussed in the previous chapter.

[‡] Where the boy studied law.

of future president John Adams before becoming a Judge Advocate in the Continental Army and finally returning to private practice. Thanks to both his alma mater and the army, he became a well-connected figure in Boston society. Monied too, for when John died in 1794, he left everything to William.

The young Frederic would want for nothing, except perhaps to find a purpose of his own. For, while William had planned for the boy to follow him to Harvard, Frederic decided that it would be a waste of time and dropped out of school at just thirteen. He would spend the next four years bumming around at Rockwood, amounting to little and showing none of the drive that would later define his life.

The turning point would come in the autumn of 1800, when William suggested he accompany his brother, John Henry, on a convalescent trip. Just two years older than Frederic, John Henry suffered from a bad knee that had rendered him an invalid. Warmer climes, it was hoped, would help him heal. The boys chose to travel to Havana, arriving in March 1801 as if to paradise itself. But it was not to remain that way. As the Caribbean spring gave way to summer, they found the heat increasingly oppressive, and John Henry's health showed no signs of improving. If anything, it was getting worse.

They sailed north. By June, they reached Charleston, South Carolina, which proved no better. Neither did

Virginia. With John Henry sickening still further,[*] they headed west, ending up in Philadelphia, where John Henry would die the following January.

By then, Frederic had been summoned home to Boston, where William had found him an unpaid job with his friends the Sullivan brothers. The Sullivans were merchants, trading sugar, tea, and spices to New England. Here Frederic was to gain the requisite experience for William to stake him in a business of his own. And, boy, did he have an idea for it! He was going to trade ice to the tropics.

At this point in his career, Frederic kept a comprehensive diary which records his early days in business. But despite the level of detail he'd commit to it, he does not record how he came up with his big plan. Nor does he write about his father's reaction. But he certainly knew the summer benefits of an icehouse thanks to the one in Rockwood's grounds, stocked each winter from a pond on the estate. And, in his diary, he records a draft of a letter he wrote to the senator Harrison Gray Otis, one of his father's business associates and a hoped-for investor:

---

[*] In his book *The Frozen Water Trade*—which may just be the definitive recent account of Tudor's life and work, and of the wider nineteenth century ice trade—Gavin Weightman suggests that John Henry may have been suffering from skeletal tuberculosis. Rarer than pulmonary tuberculosis, it spreads through contact with an infected person's blood or pus. It's also very hard to diagnose because it spreads relatively painlessly in its early stages. All quotations from Frederic Tudor's letters and diary are drawn from this source.

> Sir, In a country where at some seasons of the year the heat is almost unsupportable, where at times the common necessity of life, water, cannot be had but in a tepid state—Ice must be considered as out doing most other luxuries.[*]

He goes on:

> However absurd Sir the idea may at first appear that ice can be transported to tropical climates and preserved there during the most intemperate heats, yet for the following reasons does appear to me certain that the thing can be done and also to a profit beyond calculation.[†]

He goes on to claim, as proof of concept, that he had heard of an American sea captain, whom he doesn't name, who had shipped ice from Norway to England.

However, the Norway–Britain ice trade would not begin until 1822, some seventeen years after Tudor records his drafted letter in his diary, instigated by one William Leftwich, who would go on to open and operate ice wells in London, with his first sited in Cumberland Market in 1825. To what could Frederic possibly referring? After all,

---

[*] Weightman, *The Frozen Water Trade*, page 8.

[†] Ibid.

and however absurd Tudor's scheme may have seemed, new ideas and innovations rarely happen in isolation.

We call this concept "multiple invention," a process whereby several different people happen upon a broadly similar solution to the same problem. "Inventions and scientific discoveries tend to come in clusters, where a handful of geographically dispersed investigators stumble independently onto the very same discovery," writes Stephen Johnson in his book *How We Got to Now*.[*] "The isolated genius coming up with an idea no one else could even dream of is actually the exception, not the rule."[†]

Just because there was no formal Norway–Britain ice trade in place when Tudor was out there pitching for investment in 1805, it doesn't necessarily follow[‡] that he'd been misinformed. Commercial icehouses had begun appearing around Britain's major fishing ports in the latter half of the eighteenth century. You'd expect the fishing industry to be quick to catch on to the idea of icehouses as it slowly percolated through British society from royalty

---

[*]   Pages 58–59.

[†]   *How We Got to Now* is a fascinating history of innovation, focusing on, as the subtitle says, six inventions "that made the modern world." Its most striking takeaway is that the majority of these effects were never actually intended by inventors of the things themselves. "Innovations usually begin life with an attempt to solve a specific problem, but once they get into circulation, they end up triggering other changes that would have been extremely difficult to predict" (page 3). Frederic Tudor stands as a primary exemplar of this phenomenon.

[‡]   As has sometimes been assumed.

and the landed gentry downwards. After all, by this point its ability to keep fish fresh had been known for at least 1,500 years.* There were icehouses operating in the Scottish fishing port of Helmsdale from at least the late 1760s, from which they'd ship fresh salmon and cod to London. Further south, on the Anglo-Scottish border in Berwick-upon-Tweed, icehouses transformed the practice of boiling and salting salmon for sale, adding to the market for the fresh stuff. Their operation was not on a small scale. By 1799, the Berwick icehouses required some 7,600 cartloads of ice cut fresh from local water sources to meet the season's demand. And they would continue to be used, stocking Norwegian ice, as late as 1939.†

It wasn't just the fishing industry that demanded ice. In London, ice cream—or cream ices, as they were known at the time—was all the rage.‡ And the doyenne of London's ice cream scene was an Italian confectioner called Domenico Negri. Negri opened his first shop, known variously as "The Pineapple" or "The Pot and Pineapple," on

---

* Galen of Pergamon wrote of it in the second century C.E.

† Buxbaum, *Icehouses*, page 37.

‡ We should note here that, just because ice cream was fashionable and in demand in late eighteenth century London, it doesn't mean it was everywhere. It wasn't like everyone was popping into the Häagen-Dazs shop on Leicester Square—it was very much a luxury item for the wealthy. Let me put it this way: if you couldn't afford servants, you probably couldn't afford ice cream.

Berkeley Square in 1769.* He and his later partners, most
notably James Gunter, who eventually took over the business, were held in such high esteem that they would go on
to train many of London's successful confectioners thereafter. As Elizabeth David wrote, "A man who had trained
at Negri's, subsequently Negri and Gunter . . . had a
reference for life."† Most of those who wrote one of the
many books on confectionary and ices published in the late
eighteenth and early nineteenth centuries acknowledged
their debt to the master.

In all these books, the details on their various ice cream
freezing processes are closely guarded. While they all suggest the "ice and salt principle,"‡ they never tell you how.
And no one tells you where the ice came from. But come
it did, presumably from icehouses stocked each winter, just
like those on the commercial fishing wharves and the one
in the grounds of the Tudor's estate at Rookwood. Should
the previous winter have been mild and the ice stocks
insufficient, it had to be found elsewhere. This implies
some form of trade in ice, though not necessarily one of
the scale of Frederic's imaginings. And not necessarily the
same way around. Frederic's pitch to Senator Otis involved

---

* This date comes courtesy of the food writer and historian Elizabeth
  David in her book *The Harvest of the Cold Months*, page 310. Others
  say it opened in 1757.

† Ibid, page 311.

‡ Whereby salt is added to the ice to lower its freezing point, drawing
  heat from whatever you've placed within it to chill.

taking ice from where it was plentiful to places it didn't exist. In Europe at this time, those who needed it had to hire people to go, find it, and bring it back.

In a note written by the German science writer Johannes Beckmann some time between 1793 and 1805, when he was working on a paper about artificial cold, he describes a group of Hamburg merchants sending out a ship to Greenland to procure ice because the previous mild winter meant they didn't have enough. Greenland crops up again in *The Epicure's Almanack* of 1815, which tells us that "neither a Greenlander, nor a Highlander from the remotest part of Caledonia, nor yet a Norwegian would calculate the value of ice in Mr Gunter's cellar'"*—the Gunter in question being Robert Gunter, son of James, Domenico Negri's former apprentice and business partner.

One might think that this suggests some sort of ice harvesting operation on Greenland's frozen lakes and ponds, something akin to the cutting of it each winter in Britain, Germany, and elsewhere that took place to stock icehouses for the summer. And while this might have happened,[†] the reality we know of turns out to be more fascinating and much more dangerous. According to an article published in the Danish journal *Handels-og Industry Tidende*[‡] in 1822, an ice cargo was landed in London in

---

* Elizabeth David, *Harvest of the Cold Months*, page 326.

† There is no evidence for it.

‡ *The Trade and Industry Times.*

1816 and held in bond because the local Customs officers had no idea how to tax it. It had no loading papers, and no registered port of embarkation, for the very simple reason that it had never been near a port before. It had come from an iceberg.

In her book *Harvest of the Cold Months*, Elizabeth David quotes an account from a private letter made by an old Dane, once based in the west of Greenland, who tells her that harvesting ice from icebergs was not uncommon. In fact, he said, it would give you the purest drinking water possible. When asked of the danger of such a practice, he said, "If you know icebergs, you know which ones are going to tip round. You sail right up to the side of the iceberg and simply chop at it."*

How long this had been going on for, we simply don't know. But, thanks to Beckmann, we know it predated Frederic Tudor's search for investors. We also know, thanks to a letter from Cassandra Austen to her famous sister Jane, written from the coastal resort of Weymouth in 1804, in which she complains about the lack of ice that summer, that in fashionable circles ice was to be expected and its absence was remarkable.

There was precedent for ice shipping, just not on the scale that Frederic Tudor imagined it. He was looking for "a profit beyond calculation." To his mind, ice was the new

---

* Elizabeth David, *Harvest of the Cold Months*, page 327.

ginseng. And ginseng had made a number of Bostonians staggeringly rich.

In the years that followed the Revolutionary War, the United States' economy plunged into a depression. The thirteen colonies had been utterly reliant upon trade with Britain, which had kept them effectively cut off from other European markets since the English Civil War. Suddenly, that trade was gone and new markets were desperately needed.

Then, in 1783—coincidentally, the year Frederic Tudor was born—a Boston syndicate refitted a ship called *The Empress of China*, loaded her hold with thirty tons of ginseng, and sent her to Canton, now Guangzhou, in southern China. They had realized that, while New England had an abundance of the stuff and very little use for it, the demand in China far outstripped supply. They sold the lot and, after using some of their earnings to buy tea, silks, and porcelain to sell in Boston, they made a killing. The so-called Old China Trade was born.

However, New England didn't have an abundance of much else. It grew no sugar or cotton, unlike its southern neighbors, which meant that ships often left Boston harbor unladen when they sailed south to bring cargo back to the northeast's wealthy markets. But ice, which was essentially a free commodity, could fill those southbound holds. The ships' owners would, theoretically, be delighted to have a cargo—any cargo—instead of wasting money shipping nothing but air. All Tudor had to do was harvest and store his ice.

There was just one problem: No one would rent him a ship. In fact, Boston's shipowners were convinced that the ice would melt on board, the resulting water, sloshing around in the hold, would render vessels unstable, and they'd be lost.

Never one to be deterred, Tudor decided to buy his own. The *Favourite* cost him $4,750,* almost all the cash he had. To much mockery, she put out of Boston on February 13, 1806, her hold packed with hand-cut New England lake ice. Reporting its departure, the *Boston Gazette* quipped, "No joke. A vessel with a cargo of 80 tons of ice has cleared out from this port for Martinique. We hope this will not prove a slippery speculation."

Slippery or not, things did not go smoothly. Though the ice arrived in St. Pierre on March 5—130 tons of it, rather more than the *Boston Gazette*'s reported eighty—in "perfect condition," Tudor had nowhere to put it.

This was not through lack of planning. Frederic had sent his brother and business partner William, along with their cousin James, south to the island on November 2, 1805, with letters of introduction. Their mission was three-fold: secure the exclusive rights to sell ice on Martinique, site and build an icehouse, and drum up orders. Of these, they had only achieved the first. In their defense, James had been struck down with yellow fever and was being

---

* Around $105,000 today.

cared for by an old family friend who lived on the island, while William had set off on a Caribbean tour to tout for future business. Before he went, he left his brother a letter suggesting sites for an icehouse. But build one, he had not.

Frustratingly, we know very little about the reaction of St. Pierre's citizens to the ice's arrival beyond what Frederic tells us himself. William's letter told him that there didn't seem to be a demand for it here, and that he would be better off selling it elsewhere.* There's a story, which may be apocryphal, of one of Tudor's very first buyers returning to demand his money back because the ice had melted and he had bought, in effect, nothing.

Despite William's advice, and displaying the stubbornness that would eventually serve him well, Tudor decided to stay in Martinique to sell the ice himself, turning down an offer of $4,000 for the lot in the process. It did not go well. While he earned the distinction of making the first ice cream ever to be sold in the Caribbean, he barely covered his losses. In mitigation, he bought a cargo of sugar to sell on in Boston—a plan which ought to have seen him break even—only for the *Favourite* to be dismasted by a squall just two hours into its trip home. In all, Frederic lost more than $3,000 on the venture.† But he was not discouraged. He had proved it was possible to ferry ice

---

* In *How We Got to Now*, Steven Johnson writes: "The idea of frozen water would have been as fanciful to the residents of Martinique as an iPhone."

† Just over $69,000 today.

over 2,000 miles to the tropics, that it would not melt and sink the ship, and that it would maintain its quality when it reached its destination. The second of these was vital. With the *Favourite* lost, he would now have to rely on Boston's ship owners for hold space.

Buoyed by this proof of concept, Frederic spent the summer of 1806 building his contacts across the West Indies, appointing sales agents in Guadeloupe, St. Thomas, Barbados, and Jamaica. Shrugging off his lack of customers in Martinique, he also commissioned an icehouse in St. Pierre. Thanks to his cousin William Savage, he managed to have an icehouse built in Havana as well. It would be here that he would send his second load of cargo—180 tons of ice aboard a brig called the *Trident*—in January 1807.

In Havana, cousin William found a ready market for their product, selling some $6,000 worth of ice. He even managed to organize a return cargo of molasses. After Frederic's first year of setbacks, things were beginning to look up—only for the molasses's buyer to go bust, leaving Tudor with the bill. Suddenly, he lacked the funds to initiate his second ice shipment to St. Pierre, nor his first to Jamaica, Barbados, and Antigua, islands to which brother William had secured exclusive sales rights from the British government.

And things were about to get worse. In December 1807, at President Jefferson's request, the United States Congress passed the Embargo Act, a law designed to prevent violations of America's neutrality in the Napoleonic

Wars. It was a reaction to both the British and French navies seizing American cargoes as contraband, and to stop the British pressing American sailors into service. It was devastating to the US economy: it effectively banned all American ships from sailing to foreign ports, and remained in effect until 1809.

Frederic chose to spend this enforced trade hiatus on research. Though he had proved he could ship ice successfully to the tropics and at last found a market for it, he was losing too much cargo as it sailed south to the West Indies. He needed insulation. After some experimentation, trying first charcoal and then peat, he settled on sawdust.

Sawdust works in much the same way as Styrofoam.* It traps tiny air pockets between each wooden shard and, since air is a terrible conductor of heat, protects the ice from the warmth outside. It had another advantage over the other two materials Frederic tried: It was free, as was the ice itself. His only costs lay in cutting, transporting, and storing the ice at each end of its journey.

By the time the embargo ended on March 1, 1809, Tudor was all set.

Except for one small problem. He was about to go to jail.

Tudor had endured two years without the ability to trade. Deeply in debt, he was depending on the 1809 ice harvest to lift him out of penury. But late in the year, just

---

* Steven Johnson, *How We Got to Now*, page 47.

as his plans were about to come together, he was arrested on Boston's State Street, in full view of the city's mercantile community. Somehow, his father and friends scraped up the funds to settle enough of his debts to free him, and Frederic plowed on with his plans. This time, after so many obstacles, they came good.

In Cuba, Tudor made his first profit from ice. Enough of one, according to his diary, to attract a competitor, who turned up in Havana that May. "He made so poor a hand of it," Frederic wrote, "that after some days he threw his ice overboard and I encountered no further difficulty with my sales."

At last it seemed as though the ice trade had delivered, just when the Tudor family needed it most. But, just two years later, the United States declared war on Britain, closing the seas to merchant shipping once again.

Its name suggests that the War of 1812 lasted just one year. In fact, it raged for three. Frederic, just beginning to claw his way from debt when war broke out, found himself facing financial ruin.

"Have I not been industrious?" he wrote in 1814. "Have not many of my calculations been good? And have not all my undertakings in the eventful Ice business been attended by a villainous train of events against which no calculation could be made which have heretofore prevented success which must surely have followed if only the common chances and changes of this world had not happened against me?"

You can feel his pain. He wasn't wrong: he had done everything he could. But, in the process, he had accumulated more than $38,000 of debt—more than $700,000 today—ensuring he'd be arrested and imprisoned twice more before the seas reopened and he could start again.

Yet he would not give up. As he wrote on the cover of his diary, begun ten years before, "He who gives back at first repulse and without striking the second blow, despairs of success, has never been, is not, and never will be, a hero in love, business or war."

Moreover, despite his arrests, he had kept busy, somehow scraping together enough money to build an ice store in Boston to resupply his future ships and, dodging the sheriff's department at every turn, escaped to Havana as soon as the seas reopened. He took with him the carpenters and materials he needed to build a new and improved icehouse of his own design.

This new icehouse would differ from traditional ones in every possible way. Where previous designs had been built into the ground and lined with stone or brick, Tudor's would be made of wood. It would be double-walled, and the cavity between inner and outer walls would be insulated with sawdust. It would have a drain at its base and a vent at the top to release the latent heat created by melting ice. It was designed to keep the ice as dry as possible, in the knowledge that ice will melt faster in a cool damp place than in a dry hot one. Its design would be the prototype for all his icehouses henceforth.

Now, at last, it seemed as though Frederic's plan was coming good. But it was to be the following year—1816—that the final piece of the puzzle would fall into place. For the first time, Tudor shipped his ice within the United States, to Charleston, South Carolina. It was here that he realized that, if he priced his ice relatively low, he could turn it from a luxury item into something people would not want to live without. Not everyone agreed with him. His father, William, argued that ice should remain an exclusive product. But Frederic was not prepared to listen. As soon as the Charleston icehouse was complete and stocked, he began advertising his unique services.

The advert spelled out his goal. He priced his ice at eight and a third cents a pound, dropping to four cents a pound for big orders. This, he assured his potential customers, was the same low price he charged in northern cities.* He invites Charlestonians to buy ice "in such quantities as shall enable the proprietor of the house to continue the present price which cannot be the case unless ice is used as a necessary [sic] of life than as a luxury."

He even began selling domestic iceboxes of his own design, which would be cooled by ice supplied at a price of just $10 a month.† Unlike those nameless purveyors of ice from the Greenland Sea, Tudor was not content with supplying confectioners and tradesmen. He intended to

---

* Which presumably means Boston—he doesn't specify.

† Just under $200 today.

make people need his product. He wanted to hook the world on ice.

To do so, he made a bold choice. Instead of paying down his debts, he decided to plow his newly generated profits back into the business. By now, he was so far in arrears that, had he returned to Boston, he would have been promptly arrested. "I am afraid to go home," he wrote to Nathaniel Fellows, who ran the icehouse in Havana. Instead, he focused on restoring the trade to Martinique and launching a new trade to Savannah, Georgia, which he would open in the spring of 1819 using the Charleston model. If anything, the Savannah operation proved even more successful than any he had tried before. But it is in Martinique that we see most clearly the Tudor philosophy—to make ice an everyday experience—laid out.

The Martinique icehouse was to be run by Stephen Cabot, scion of one of Boston's most prominent families and a man in need of a job. Once the sales rights had been resecured and the cash raised for his passage, Cabot sailed for St. Pierre, arriving in the autumn of 1818 and building the new icehouse in time to receive supplies in March. The ice sold well. So well that, in July, the Martinique icehouse ran out. To compound the problem, there was no ice available in Boston to resupply him, so Cabot was forced to take alternative action: he hired a whaler from Maine—called Captain Hadlock—to bring him back an iceberg.

Elizabeth David's Danish correspondent was remarkably cavalier about the dangers of harvesting an iceberg.

But in his book *Journal of a Voyage to the Northern Whale Fishery*, Captain William Scoresby, Jr. leaves his readers in no doubt that any dealings with icebergs were extremely perilous, describing Greenland voyages as the most arduous of maritime adventures. He writes:

> In the summer of 1821 the captain of a whaler that had been wrecked in Baffin's Bay, wishing to make himself useful in the ship that he had fled for refuge, offered to assist in fixing an anchor to an iceberg, to which it was expedient that the ship be made fast. He was accompanied by a sailor to the berg, and began to make a hole for the reception of the ice-anchor: but almost the first blow that he struck with the axe occasioned an instantaneous rent of the mass of ice through the middle, and the two portions fell in opposite directions. The captain, aware of his danger, the instant the ice began to move the division on which he was situated, in the contrary direction of its revolution, and fortunately succeeded in balancing himself on the changeable summit until it attained an equilibrium. But his companion fell between the two masses, and would probably have been instantly crushed or suffocated, had not the efflux of water, produced by the rising of the submerged parts of ice, hurried him from between them,

almost alongside of the boat that was waiting
near the place.*

So it proved for Hadlock, whose crew hand cut the
ice with picks and crowbars on a surface that pitched and
rolled beneath them, aware all the while that any shift in
the iceberg's weight distribution could flip it over and rip
a hole in the ship. Which it did.

But despite all, Hadlock succeeded in his mission and,
resupplied at considerable risk, Cabot's icehouse was up
and running once more, and doing seemingly so well that
he became convinced that he was doing better in St. Pierre
than he actually was. In addition to petitioning Tudor for a
bigger share of the business,† Cabot was spending his way
into debt. Worse for Frederic, he wasn't following Tudor's
updated marketing plan. It would have to be explained
once more.

"The creams are the first great object," he wrote to
Cabot, "and thereafter cold drinks sold at the same price
as warm. A man who has drank [sic] his drinks cold at the
same expense for one week can never be presented them
warm again"—a point Frederic had proved thrice over,
offering bartenders a free ice consignment and asking them
to do just that. "Once we have persuaded 100 persons by

---

* William Scoresby, Jr., *Journal of a Voyage to the Northern Whale
  Fishery*, pages 300–301.

† Which he didn't get.

means of our same price plan, these 100 will soon carry with them 100 more and the ratio will compound faster than we can calculate."

Tudor knew what he was talking about: he had now been trading ice for thirteen years and, in Savannah, the plan was working. Only one sixth of his stock there was bought by the rich. The rest of it was going to tradesmen—butchers, dairies, barkeeps—and citizens from all walks of life. But, to his frustration, Cabot wouldn't listen, continuing with the outdated business model and draining the profits from Charleston and Savannah, profits Tudor needed desperately to expand operations to New Orleans.

Help would come in the form of his elder brother, William, the cofounder and editor of the *North American Review*, whose work had gained him enough of a following to gather together a group of friends and fans as an investment consortium for the New Orleans expansion. The deal was that they'd provide the funds if Frederic paid William an annual annuity to keep writing. Tudor agreed. Now he had to move fast to make sure everything was in place to begin shipping with the February ice harvest. Most importantly, he needed an agent there to manage his affairs—someone with a little more nous than Stephen Cabot.

Younger brother Harry proved to be the man.

In his youth, Harry had been something of a wastrel. But even though he was in debt and in need of a way out of

Boston, Frederic was convinced he had done some growing up and was ready for the task. And so he was: on arrival in New Orleans, Harry sited and built the icehouse and even bought an ice cargo that arrived from Marblehead, north of Boston, before the Tudors' ice made port.

But despite Harry's show of promise, Frederic was making himself sick with worry. No wonder: Cabot's Martinique part of the business failed in 1821, costing Tudor some $19,000.* The last two Havana icehouse keepers had died. The business was in trouble. Struck down by a fever, all he could do was write, in block capitals, the word ANXIETY across different days in his diary. It was to prove the beginning of a full-on nervous breakdown.

Then, on July 12, 1821, he received a letter from Harry telling him that the New Orleans icehouse was selling four times what they'd anticipated, and the cash was rolling in. New Orleans saved everything. At last, Frederic was able to step back from the business, leaving it in the capable hands of his friend, Robert Gardiner, while he convalesced. As Tudor wrote to his mother, "I cannot say what would have become of me but for his stepping in to relieve me at this time."

In 1821, Tudor's business remained centered on the United States and the Caribbean. But we begin to notice the ice trade developing elsewhere, notably in London. That same year, William Leftwich began his business

---

* About $472,000 today.

importing ice from Norway. Whether he knew of and was inspired by Tudor, we don't know. But the demand, as we've seen, was there. One of its more obscure examples appears in the shipping records from Lloyd's List, which records the landing of a cargo of ice at the London docks by a Captain Smith of the *Platoff* on June 29.

That we know of it at all is down to some exhaustive research by Elizabeth David, provoked by finding the following announcement, placed in the *Times* by Robert Gunter[*] on July 5:

> Messrs. Gunter respectfully beg
> to inform the Nobility and those
> who honour them with the commands
> that, having this day received one of
> their cargoes of ice by the Platoff,
> from the Greenland seas,
> they are able to supply their
> CREAM and FRUIT Ic.e.S at their
> Former prices.—7 Berkley Square.

Neither the Negri archives nor those of the ship's owners, WH Hobbs and Sons,[†] tell us who instigated the

---

[*]   Of the confectioners Negri and Gunter.

[†]   WH Hobbs and Sons were ship owners and lightermen to the Russia Company, formerly known as the Muscovy Company, which had made the first serious attempts to open a whaling fishery in the Arctic Ocean during the reign of King James I.

Platoff's voyage north. We don't know if Robert Gunter commissioned the expedition, much like Stephen Cabot hired Captain Hadlock four years before. And we don't know if ice harvesting was viewed as a money-spinning enterprise.

What is clear is that Frederic Tudor's vision was fundamentally different to any ice trade that had gone before. In London, it was very much a business-to-business proposition. Confectioners and fishmongers needed ice, and there was a supply to meet their demands. In fact, in Britain at least, ice for fishmongers was duty-free. Tudor was out to change people's behavior: he wanted his consumers to need the luxury of ice. To do that, he needed to educate his market.

"Ice," he later wrote, "when first introduced to a population unaccustomed to it, is a nine days [*sic*] wonder and after that is over, they will very few of them sit about to use an article to which they are totally unaccustomed and to most of them is a mere sight of something which they have never seen before." It wasn't enough to sell the ice, you had to teach people how to use it as well.

To do so, Tudor gave customers insulated ice containers he had designed himself. "They pay for themselves in one year and last as many years as are taken care of," he wrote to one of his agents, who viewed them as a waste of money. To his mind, the more the customer understood what they were buying and what it could do for them, the more they would realize that ice was not a frivolous and short-lived luxury, but something of necessary, affordable, and long-term use.

It was one thing to create the demand. It would be quite another to sustain a supply of sufficient quality that he could, in words attributed to the Tesco's founder Jack Cohen nearly a hundred years later, stack it high and sell it cheap. It was to this that he turned his attention next.

Cutting ice was hard work. It had to be done by hand. The cutter had to break large pieces from a thick ice sheet with a pickaxe before cutting each piece into manageable blocks with a two-man saw. Or you could cut a hole in the ice and saw out the blocks directly from the frozen surface. Either way, it was backbreaking.

All this changed in 1825 when Nathaniel Wyeth invented the ice plow.

Nathaniel lived and worked at the Fresh Pond Hotel in Cambridge, Massachusetts, which was owned by his father and stood, unsurprisingly, on the shore of Fresh Pond.* Given that the hotel had its own icehouse, stocked from the pond each winter, he had a working knowledge of ice. At some point in the early 1820s, he began supplying Fresh Pond ice to Frederic Tudor to make some extra cash.

---

* If you're unfamiliar with Fresh Pond, it's probably best to try to scrub any images of ponds from your mind. Think natural reservoir or small lake instead. Fresh Pond boasts over three kilometres of shoreline and a sixty-three hectare surface area.

After what may have been one or several seasons of cutting ice by hand—unlike Tudor, Wyeth didn't keep a diary, so we don't know—Nathaniel hit upon his ice plow. Drawn by two spike-shot horses, the ice plow both cut into the ice and scratched out a parallel line beside the cut to show where the next one should be made. Once the harvest area had been marked out, repeated plow passes deepened each cut to the required depth, then ice blocks could be chiseled free. Not only did this speed the harvesting process significantly, it had the added virtue of creating blocks with a regular shape. They could now be stacked more efficiently both in icehouses and in ships' holds, which ensured they would melt more slowly.

Tudor was impressed. In November 1826, he offered Wyeth $500* a year to run the cutting, storing, and loading of ice for the entire business. Wyeth had revolutionized ice harvesting just as Tudor had revolutionized its insulation and transport. Assuming the winters to come were not unduly mild, he had made sure that Frederic Tudor's supply would never run out.

Even so, Tudor would not expand the range of his business for another seven years when, in September 1833, a brig called the *Tuscany* brought a cargo of his ice to Calcutta. The following year another brig, the *Madagascar*, delivered the same to Rio de Janeiro. Just two years later,

---

* Just over $14,000 today.

after thirty years in the trade, Frederic wrote in his diary, "Last year I shipped upwards to 30 cargoes of ice and as much as 40 were shipped by other persons." What was once a matter of ridicule was now big business. By 1850, more than 100,000 tons of New England ice was shipped around the world each year, more than half of it by Tudor's competitors.

And, boy, were there competitors! The Knickerbocker Ice Company was founded on Rockland Lake in New York State in 1831, a location that would later be described as the icehouse of New York City, where they would soon be consuming more than 285,000 tons of ice a year. The Wenham Lake Ice Company opened in Wenham, Massachusetts in 1842 and would claim in its advertising that its ice was not only purer than anyone else's, but that it would make things colder. These two were just the tip of the ice-trading-berg. By the end of the century, there would be so many companies that the ice business required its own trade journal. The *Cold Storage and Ice Trade Journal* would be published twice a year until the First World War. Each of these outfits exported ice from in and around New England to the waiting world. Frederic Tudor had shown them how.

From shipping to storage to retail, Tudor had built a market for others to exploit. Thanks to his canny bar-based marketing strategy, he had fundamentally changed the way America drank, to the extent that "cold" became a defining point of difference between particularly

American and British drinking preferences.* In fact, one could argue that the designation of an "American" bar, like the one that opened at London's Savoy Hotel in 1889, is as much a signifier of its prodigious use of ice as it is of serving cocktails.

This is not to say that the British didn't want ice; rather that the kind of market Tudor created for it in the United States, the Caribbean, and India never caught on there. The Wenham Ice Company attempted it in 1845, opening a shop on The Strand from which they sold iceboxes for domestic use and making ice deliveries twice a day. In addition to giving iceboxes and delivery subscriptions to Queen Victoria and Prince Albert, they also placed a block of their famously clear ice in their shop window with a copy of that day's *Times* behind it. People would queue up to read the news through it. But, while they would open stores in Manchester, Liverpool, Birmingham, Hastings, and Dublin, their advertising made a deeper impression on the public consciousness than their product—to the extent that Norwegian exporters petitioned the Norwegian Crown to rename Lake Oppegård to Wenham Lake so that they could legally sell "Wenham Lake" ice in London—and

---

* There's a telling beat in Peter Weir's 1982 movie *The Year of Living Dangerously*, set in Jakarta in 1965, in which the stuffy British Colonel Henderson berates a waiter, saying: "This is not a gin and tonic. This is gin, tonic, and ice. Americans always have ice, and I am *not* an American." Anecdotally, when I was a little boy in the 1970s, my mother would only allow us one ice cube per glass of water to chill it on a hot summer's day, despite being raised in Jamaica.

the company failed. It was bought out and refocused on the US market in 1850.

The British market was too small for the Americans. While the Norwegians were exporting 900,000 tons of ice a year to Britain by 1900, the Knickerbocker Ice Company was shifting four million tons of it alone.

Ice use had now moved far beyond the bars and ice cream parlors of Tudor's early marketing plans. The domestic ice subscriptions he pioneered in Charleston and Savannah, icebox included, made ice deliveries an everyday feature of normal life for a middle-class that was increasingly affluent. They had become so used to it that they found it hard to imagine life without it. Such was the demand that, by 1860, two out of three New York households enjoyed daily ice deliveries.

In India, too, customers became dependent on Tudor's ice. When supplies in Bombay ran out in July 1850, the city's *Telegraph and Courier* announced, "We are without ice! The supply is used up," before going on to say that citizens should agitate in the streets against "the abominably sudden and inexplicable cutting off of our supply of ice." We can catch a glimpse of just how indispensable iceboxes had become to those who had grown up with them in Gerald Durrell's *My Family and Other Animals*. Shortly after arriving in Corfu, Louisa Durrell recalls the American iceboxes of her Indian childhood and commissions something similar to be made, all the better to survive those long hot summers. Such an icebox would have been incredibly rare in the Corfiot kitchens of the time, yet there was ice available to chill it.

Single-handedly, Frederic Tudor had revealed that cheap ice was a holy grail the world didn't know it needed. But he could never have anticipated the effects his product's ubiquity would have upon industrialized societies, and upon the United States in particular.* Cheap ice and the technologies that rode in upon its coattails would change the way we ate and accessed our food, the way we practiced medicine, and even shift entire demographics across continents.

Whether he thought of it like this or not, Frederic Tudor had inadvertently invented the cold chain. Though he was never to use his ice to ship foodstuffs profitably, others would. Railway cold chains would radically change the food supply of the United States in the second half of the nineteenth century. The meatpacking industry, from its hub in Chicago, would lead the way.

Three factors combined to transform meatpacking from a regional business into a national concern. The first, the American Civil War, ensured huge supply contracts that allowed entrepreneurs to recognize and embrace economies of scale in the trade. Second, the railways allowed meat to be sent over long distances quickly enough for it not to spoil. Finally, ice guaranteed

---

* In *The Frozen Water Trade*, Gavin Weightman describes America, even as early as the 1840s, as "the first refrigerated society."

freshness over time, just as it had with the Berwick salmon trade at the beginning of the century.

From 1875, specialized ice-cooled rail trucks shipped meat and vegetables across the United States with replaceable ice blocks suspended in bins above the produce on board. By the 1880s, the fresh food business was one of the largest consumers of harvested ice in the country.

Meanwhile, thanks in part to ice, the railroads' rapid push into western states and towards California in the years following the Civil War encouraged a demographic shift from east to west. It would be aided by advertising proclaiming California "the cornucopia of the world" with "room for millions of immigrants"—all this farmland, whose produce could be sold back east thanks to ice-chilled trucks.

Ice would change rail travel for passengers, too. In 1878, the restaurateur Fred Harvey signed a contract with the Atchison, Topeka and Santa Fe Railway to open up restaurants along their lines. Prior to this, railway dining in the west was a perilous affair. Back east, with Pullman cars and relatively short distances between destinations, one could count on a small measure of luxury. Out west, with its hundreds of miles of deserts and prairie, you'd be lucky if the coffee was less than a week old, luckier still if your food order arrived before your train was ready to depart, and luckiest of all if it was freshly cooked.

Harvey changed all that. Ice-cooled rail cars allowed him to transport and serve ingredients like Montana brook trout and New England oysters in the heart of New Mexico. Thanks to his policy of hiring single, educated women from the East Coast and Midwest between the ages of eighteen and thirty to staff his Harvey Houses, he kept, in the words of the cowboy and movie star Will Rogers, the west supplied with food and wives.* While the railway opened the southwestern United States, Fred Harvey civilized it. And it would not have been possible without ice.†

As Frederic Tudor came to the end of his life, he could reflect with pride on his accomplishments. He had created a trade that shipped something that once was worthless to the world—beyond India and the Caribbean, New England ice now traveled to Manilla, Singapore, Hong Kong, Guangzhou, and Australia—and it had made him rich beyond his dreams. By the time of his death in 1864, the North American trade in lake and river ice was just a fraction of what it would become. A report commissioned by the US Census in 1880 and published in 1883 estimated

---

* The full quote reads: "In the early days, the traveler fed on the buffalo. For doing so, the buffalo got his picture on the nickel. Well, Fred Harvey should have his picture on one side of the dime, and one of his waitresses with her arms full of delicious ham and eggs on the other side, 'cause they have kept the West supplied with food and wives."—cf. Lesley Poling-Kempes, *The Harvey Girls*, published by Paragon House, New York, 1989, page 102.

† Once again, as Allie Fox observes in *The Mosquito Coast*: "ice *is* civilisation."

that as much was spent on ice year-round as was spent on winter fuel, so great was the demand.

Ice could be harvested so cheaply that artificial, plant-manufactured ice could not compete. It was simply too expensive to make, and the first commercial freezers couldn't meet the demand. Even as late as 1906, the *New York Times* reported, amid fears of a mild winter and resulting ice famine, that the city needed four million tons of it a year, and the ice plants could produce as little as 700,000. Ironically, it would be the very industrialized development that exploded alongside the ice trade—helped along in no small part by the improvements in food standards that ice had helped to nurture—that would kill the practice of ice harvest in favor of machines.

In 1907, Dr. Daniel D. Jackson, a chemist working for New York's Department for Water Supply, Gas, and Electricity, wrote a report saying Hudson River ice was unsafe for human consumption. By the outbreak of the First World War, the natural ice business was all but done for. Plant ice would prevail hereon. The stage was set for the rise of the refrigerator.

4

The Convenience of Cool

I don't care," Frank Sinatra sang to Grace Kelly,[*] "if you are called \ The fair Miss Frigidaire \ 'Cause you're sensational!" If you think about it, the fridge is sensational as well, perhaps more so than Miss Kelly.

Food writer Bee Wilson has described it as a box of desires, the hearth around which the modern kitchen is built.[†] Science writer Tom Jackson says it's the fridge that makes the modern city, before offering the staggering statistic that those in Greater Tokyo, "the world's largest urban area, provide ingredients for at least 113 million meals a day."[‡]

That I can walk out of my office right now, on a sweltering 33°C (91.5°F) day, go into the kitchen, and fill

---

[*] In *High Society.*

[†] The *Daily Telegraph,* July 3, 2011.

[‡] Tom Jackson, *Chilled*, page 10.

my water glass with fresh ice straight from my freezer is nothing short of a modern miracle.

Like those mid-nineteenth century bartenders with their ice in the previous chapter, it is something we take utterly for granted. And yet, together with its close cousin, the air conditioner, the refrigerator has transformed the way we live, making cool the ultimate convenience. It is to their stories, and those of the people behind them, that we now turn.

By the 1880s, the citizens of the United States were consuming over five million tons of natural ice a year, a staggering figure that brought with it a very nasty problem: disease. Most particularly, dysentery and typhoid. Many sources of natural ice were no longer as pristine as once they were.

The lakes of New England had become famed for their clean water and clear ice. But close inspection of said ice would reveal a hint of green caused by tiny plant particles and algae. Moreover, a harvesting practice known as "sinking the pond," whereby you drilled holes into the ice to cause water to well up through them, the best way to create more ice, added further impurities by trapping anything trodden or blown onto the surface beneath the newly frozen layer. No matter how closely you followed Jerry Thomas's ice cleaning advice, the dirt was trapped within.

The real problem was not flecks of dirt, but microbes. The demand for ice was such that the lake surface coverage in New England was proving insufficient to meet it, and

harvesters turned to rivers which, thanks to the twin benefits of industry and sewage, were becoming increasingly
polluted. On the Schuylkill River, an ice source for the
Knickerbocker Ice Company of Philadelphia, pollution
was not only killing the fish, it turned the ice dark green.*
Even on St. Patrick's Day, green ice is far from appealing,
and the Knickerbocker Company ended up importing
cleaner ice from Maine.

Natural ice would effectively be done for just twenty
years later after a typhoid outbreak at a mental hospital in
Ogdenberg, New York, in 1902. It turned out that their
ice supply harvested from the St. Lawrence River came
from a reach fed by the hospital's own sewer.†

The time was ripe for an alternative supply, and the
technology had already been invented to make it. The only
reason it hadn't fully replaced natural ice already was that
it was cumbersome and more expensive, and natural ice
suppliers had spent quite a lot of time and money per-
suading people that their ice was cleaner than something
made in a machine.

The first documented account of manufactured ice
dates from a 1756 scientific experiment conducted by
William Cullen at the University of Edinburgh. Together
with his pupil, Matthew Dobson, Cullen was at the time
exploring the cooling effects of evaporation. Dobson had

---

* Tom Jackson, *Chilled*, page 162.

† Ibid, page 164.

noted that, whenever he took a thermometer out of alcohol, it always dropped by a few degrees, and he suggested that this cooling was caused by its evaporation from its surface. As they experimented further to get to the bottom of this, Cullen noticed that the thermometer also dropped when it was placed in a vacuum chamber and the air was sucked out. This effect proved even more remarkable when the thermometer in the vacuum chamber had been dipped in alcohol first.

Cullen theorized that the vacuum was exacerbating the rate of evaporation, and therefore its effects. Exploring further, he filled a vessel with water and placed a smaller one filled with ethyl nitrate inside it. He then put these in the vacuum chamber and pumped out the air. The ethyl nitrate evaporation was so pronounced that the water froze.[*] Cullen didn't take this research any further. In fact he didn't even publish it in full. Other scientists would step forward to unravel the mysteries of temperature, among them William Thomson, later Lord Kelvin,[†] and James Joule.[‡]

Among Joule's many accomplishments was measuring the mechanical equivalent of heat, an important step in our understanding that heat is a form of energy, and cold the absence of the same, which underpin the later laws

---

[*] Tom Jackson, *Chilled*, pages 80–81.

[†] For whom the kelvin temperature scale is named.

[‡] The joule, as a unit of energy, is named after him.

of thermodynamics. This achievement would fascinate Thomson and lead the two to conduct research together, by letter, from the mid-1850s on. In 1852, they would discover what is known as the Joule–Thomson expansion. You take two flasks connected by a very narrow tube. You fill one with a gas, and you pump everything out of the other to create a vacuum. Then you pump the gas into the vacuum. As the gas expands to fill it, the temperature drops. It is this principle that makes a refrigerator work.[*]

While Thomson and Joule may have codified the science, they did not invent the fridge. The men who would had it figured out before the science was fully understood. American inventor Jacob Perkins filed the first patent for a vapor-compression refrigeration system in both England and Scotland on August 14, 1834. In this, he was drawing upon the work of another American inventor, Oliver Evans, a steam engine pioneer and developer of the automatic flour mill. Evans created his design in 1805 but never built it, even though it is the first to understand the principles Joule and Thomson would go on to prove.

Perkins actually built his in London, writing, "I am enabled to use volatile fluids[†] for the purpose of producing the cooling or freezing of fluids, and yet at the same time constantly condensing such volatile fluids, and bringing

---

[*]   Tom Jackson, *Chilled*, pages 126–129.

[†]   In this case, ether.

them again into operation without waste." It worked, but no one wanted to buy one.

The second was a Florida-based American doctor named John Gorrie in 1842. Gorrie worked at a naval hospital in Apalachicola, where he specialized in tropical diseases, particularly yellow fever and malaria. At the time, no one knew that both of these were transmitted by mosquitoes. Gorrie had noted that the majority of cases tended to appear in the summer, which fitted the theory that they were spread on the "bad" hot summer air from which malaria gets its name. He quickly latched onto ice's potential to help his patients, suspending large blocks of it from the ceiling so cooler air could flow down onto them.

This early attempt at air conditioning seems to have worked. But the ice supply was intermittent, and Gorrie sought a more permanent solution: a machine.

Unlike Perkins's ether-based version, Gorrie relied on air. Describing it in the pages of the Apalachicola *Commercial Advertiser* in 1844, he wrote, "If the air were highly compressed, it would heat up by the energy of compression. If this compressed air were then run through metal pipes cooled with water, and if this air cooled down to the water temperature was expanded down to atmospheric pressure again, very low temperatures could be obtained, even low enough to freeze water in pans in the refrigerator box.""

---

* Gavin Weightman, *The Frozen Water Trade*, page 165.

It worked. And Gorrie did it entirely on his own. That it's possible for people in very different parts of the world to happen upon the same technological solution to a problem comes down to what the complexity theorist Stuart Kauffman has dubbed "the adjacent possible."* In his book *How We Got to Now*, the writer Steven Johnson argues that this is because new ideas are formed from networks of other ideas that already exist. "We take the tools and metaphors and concepts and scientific understanding of our time, and we remix them into something new. But if you don't have the right building blocks, you can't make the breakthrough, no matter how brilliant you are."† Enough ideas had come together to allow Evans, Perkins, and Gorrie to make mechanical refrigeration possible, and Joule and Thomson to nail down the physical science of it, independently of each other.

Gorrie believed his device would be so successful that he gave up his medical practice to pursue it further, securing his patent for it on May 6, 1851, and setting up business in New Orleans. But it was not to be. It didn't help that his business partner died, nor that Frederic Tudor began taking out advertising to slander the cleanliness of his artificial competitor's product. But what really spoiled his dreams was that his machine was not up to the task. New Orleans was by now consuming so much

---

* Steven Johnson, *How We Got to Now*, page 36.

† Ibid.

ice that Gorrie's contribution would be little more than a drop in the ocean.

Gorrie would die broken and penniless in 1855, but the idea of artificial cold would not. In the same year that Gorrie received his patent, a British émigré to Australia, James Harrison, built his first ice machine in Geelong, Victoria, where he owned the local newspaper. His first commercial machine, using an ether compression system, appeared just three years later. And it was huge—you would struggle to fit it into the average house. But it worked, and it sold. Harrison would go on to fit them into cargo ships, allowing fresh produce to be sold in foreign markets.*

Harrison was not alone in creating new systems independently, all of which worked on the same compression principles. In France, Ferdinand Carré developed the first gas absorption refrigeration system in 1859, working from previous experiments conducted by his brother Edmond.

It worked by dissolving gaseous ammonia in water. When you heat that water, the ammonia evaporates, creating cooling. As the gas cools, it is then reabsorbed into the water before the solution goes back to the heating unit for the process to begin again.

It's a system that has a couple of advantages over anything that came before. For one, it has no moving parts.

---

*   They would also turn out to be very popular with the Australian brewing industry, something that would ultimately give rise to the expression, "Crack me open a cold one, will ya."

And two, because of the ammonia's low boiling point, it doesn't require a lot of heat to make it work.

Meanwhile, in Munich, the German engineering professor Carl von Linde was hard at work on the liquification of gases, patenting his system in 1876. It was work which would require him to develop a substantial understanding of refrigeration. So substantial, that his knowledge led him to give up his professorship to found the *Gesellschaft für Lindes Eismaschinen Aktiengesellschaft*, or Linde's Ice Machine Company, in 1879. As with Harrison's ice machines in Australia, Bavaria's breweries were particularly keen on Linde's work. So too were slaughterhouses across Europe. His machines would form the basis of the first cold storage operations on the continent. Returning to academia, Linde would go on to design a system for the liquification of carbon dioxide for the Guinness Brewery in Dublin and work on the fractal distillation of liquid aid to obtain pure oxygen and nitrogen.

His legacy in the industry can still be felt today. His work led directly to the creation of stable ammonia refrigeration cycles, with ammonia still being used as an industrial refrigerant today. And his entrepreneurial streak saw him secure himself a stake in Brin's Oxygen Company, later the BOC Group, in 1906, and he remained on its board until the outbreak of the First World War. In 2005, Linde plc, the multinational chemical conglomerate his original company had grown into, would go on to buy BOC outright.

So far, so industrial. While fishermen, slaughterhouses, dairies, and breweries now benefited from the reliable clean ice these machines supplied, little had changed for ordinary people. Their cooling still worked much as it had when they were supplied by the natural ice magnates of New England and Norway. The self-contained fridge would not arrive until the second decade of the twentieth century.

The fridge as we understand it today would be invented by Fred W. Wolf in 1913. Or by Nathaniel B. Wales in 1914. Or by Alfred Mellowes in 1916.

Wolf's Domelre, a name composed of the first syllables of the words "domestic electric refrigerator," had a number of pioneering features, not least the fact that it was electric. It had a thermostat to control its temperature. It had an air-cooled condenser. And it had the first ever ice cube trays.* Despite a price tag of $900,† Wolf sold hundreds, if not thousands, of them,‡ and would go on to sell his business to Henry Joy, then the president of the Packard Motor Company. However, despite improvements to the refrigerator's efficacy and price reductions, the business did not prosper under Joy's ownership and folded in 1922.

---

* Bernard Nagengast, "It's a Cool Story," *Mechanical Engineering* 122, 5 (May 2000): pages 56–63.

† About $26,400 today.

‡ Estimates vary.

Wales and Mellowes fared considerably better. Wales's company, based in Detroit and backed by another car manufacturer, this time Arnold Goss of Buick, was cumbersomely called the Electro-Automatic Refrigerating Company when it launched in 1914. Perhaps unsurprisingly, he renamed it two years later in honor of William Thomson, now dubbed Lord Kelvin, and the Kelvinator was born. It was wildly successful. By 1923, the Kelvinator Company held an 80% market share for electric fridges in America, and they would expand into the British market three years later.

Mellowes brought his fridge—the first to have its compressor on the bottom of the cabinet—to market in 1916, together with his partner J. W. Murray. And, while he was not as initially successful as Wales, his work came to the attention of William C. Durant, the cofounder of General Motors. Durant bought the business and renamed it Frigidaire, a brand name that would go on to be as synonymous with fridges in some parts of the world as Hoover became with vacuum cleaners.

But* there was a problem. These devices' refrigerants used chemicals like sulfur dioxide, which is highly poisonous, and methyl formate, which, while also toxic, quite likes to explode.

It is hard to find reliable statistics for the number of people these devices killed. But it was enough of a problem

---

* There's always a "but" in these stories.

for Frigidaire to seek a solution. And not only them. One morning in 1926, the great Albert Einstein was shocked to read in the newspaper that an entire Berlin family had died in their sleep when their refrigerator's compressor seal had failed and their apartment filled with toxic gas.

"There must be a better way," he is reported to have said to his colleague and friend Leo Szilard.

The first Einstein–Szilard fridge uses three gases in its system rather than two. It requires no electricity to run, which means it has no motor and operates silently, and does not require seals to retain any toxic gases. They would go on to design two more—in one design, they came up with an ingenious pump that operated with molten sodium; in the other, they designed a vacuum pump that would be powered by the water pressure from the tap—but none of them made it to market. A more convenient solution to the problem had been found by a team of General Motors' scientists that included a man named Thomas Midgley, Jr.

It is probably fair to say that Midgley is responsible for more death and human suffering than anyone who hasn't started a war. But it is only nowadays that we view Midgley as a bad guy. In his day, he was a revered and award-winning chemist, celebrated for having solved two of the modern world's mechanical problems.

As would seem appropriate for a man who went to work for General Motors in 1916, when he was twenty-seven, the first of these had to do with cars, specifically an issue called engine knocking. This is a phenomenon where, when the fuel and air are fed into an engine's cylinders, it does not combust at the correct time in the cylinder's stroke.* It sounds like the cylinders are being tapped by a hammer. You can still experience knocking in a modern engine,† but the cause will almost certainly not be the same as it was in the 1910s and '20s. Back then, it was the result of poor petrol quality. Left untreated, knocking will slowly erode the piston head, ultimately breaking the whole damn thing. Which, in the early days of aviation, would result in plummeting and death.

Midgley was charged by his boss at the Dayton Research Laboratory,‡ Charles Kettering, with cracking the puzzle. After trying a host of different additives, Midgley hit upon ethanol, duly patenting the formula for an antiknock fuel in February 1920. But General Motors boss Alfred P. Sloan didn't want it. For a start, you couldn't patent ethanol itself—anyone could make it. In fact, with Prohibition now the law of the land, "anyone" pretty much was making the stuff themselves if they fancied a drink. Which meant that "anyone" could buy their fuel untreated

---

* It is not to be confused with a misfire, which happens when one or more spark plugs fail to fire.

† And if you do, you should get it fixed immediately, because it will create an expensive problem for you.

‡ Owned by GM.

and mix their own antiknock fuel, and GM would not make the profit Sloan wanted off the innovation.

So Midgley was sent back to the lab, where he hit upon a new and deadly solution: tetraethyl lead (TEL).

As far as GM was concerned, this was absolutely perfect. Even though TEL had first been discovered in 1853 by a German chemist named Karl Jacob Löwig, it had until now no commercial usage. It was cheap to make. The mixture could be patented. And GM would make a fortune off it.

Fuel-caused knocking was solved. Under the brand name Ethyl, leaded petrol went on sale in 1923. Meanwhile, Midgley went to Miami on vacation to cure himself of lead poisoning.

At no point was the word lead mentioned in any of the new product's advertising.

Then people started dying. In October 1924, "workers in a Standard refinery in Bayway, New Jersey, went violently insane after making leaded gasoline. Seven men died and 33 were hospitalized there; meanwhile ten more were killed at a DuPont factory, and at least two died and 40 were hospitalized in Dayton, Ohio."* After a federal inquiry, the Public Health Service decided that the lead in petrol posed no threat to public health.

---

* William Kovarik. "Ethyl-leaded gasoline: How a classic occupational disease became an international public health disaster," *International Journal of Occupational and Environmental Health*. 11 (4) (2005), pages 384–397.

They could not have been more wrong. Midgley must have known it, too, despite being trotted out at a press conference on October 30, 1924, where he argued for the product's safety by pouring the stuff all over his hands before wiping it off with his handkerchief and then sniffing the bottle for a full minute. No one was poisoned, he insisted, and any injuries had been "caused by the heedlessness of workers in failing to follow instructions."[*]

By the time leaded petrol was finally banned,[†] it would be impossible to estimate how many people had been killed, directly or indirectly, by exposure to leaded petrol or the toxic gases its burning released into the atmosphere. But its phasing out and banning from the 1970s on has not only led to cleaner air—it has also led to the hypothesis that the decrease in atmospheric lead resulted in lower crime rates from the 1990s on.

Midgley was not yet done inventing pollutants. Having solved engine knocking for GM, setting them on a path to billion-dollar profits with the work, he was tasked with solving the issue of toxic refrigerators, once again at Kettering's behest.

Midgley and his team were looking for a chemical that was highly volatile, a refrigeration requirement, nontoxic,

---

[*] Ibid.

[†] With Colombia leading the charge in 1991, and the Yemen bringing up the rear in 2018.

odorless, and completely inert. After some experimenting with chemicals called alkyl halides,* they decided to work on a concept that would combine fluorine with hydrocarbons. In 1930, they succeeded, synthesizing dichlorodifluoromethane, which would come to be known to the world as Freon, for the first time. Over the next five years they would go on to develop four more chlorofluorocarbons, believing these compounds to be completely inert, as they had been tasked.

Midgley demonstrated Freon's nontoxicity and stability to the American Chemical Society by inhaling a lungful of it and then blowing out a candle. Unpoisoned and unblown up, he would go on to be awarded the Society of Chemical Industry's Perkin Medal for its creation.

Freon would quickly replace all other refrigerants in domestic fridges and freezers. The wider family of CFCs would go on to be used in aerosol propellants and degreasing solvents around the world.

While Midgley almost certainly knew the dangers of the lead he'd put in petrol, having poisoned himself with it during his research, there is no way he could have known the effects his CFCs would have upon the upper atmosphere, where the chemicals would prove to be anything but inert, ripping apart ozone atoms that bounce back

---

*   These are compounds where hydrogen atoms are combined with a halogen atom—fluorine, iodine, chlorine, or bromine.

ultraviolet radiation into space and releasing their pollutive base atoms—chlorine and fluorine—into the sky.

It is estimated that the Montreal Protocol, which banned CFCs in 1989,* could end up averting 443 million cases of skin cancer, 2.3 million deaths caused by those cases, and 63 million cataracts cases for people born between 1890 and 2100 in the United States alone.† Hard statistics of a death count that we can attribute to Midgley's work are hard to come by—apart from his own. In 1940, he contracted polio, which left him paralyzed. So he invented a system of ropes and pullies to help him get out of bed. With this *Wallace and Gromit*–like contraption, he managed to strangle himself accidentally in 1944.

Midgley's work on CFCs, however, led to a meteoric rise in American fridge sales, despite the country being deep in the grip of the Great Depression. At the beginning of the 1930s, just 8% of US households had one; by the decade's end, that had risen to 44% and, by the end of the 1950s, to a staggering 96%. Indeed, as the writer Matt Novak puts it,

> The refrigerator came to be one of the most
> important symbols of middle-class living in the

---

* There's more on this in chapter 8.

† https://ozone.unep.org/montreal-protocol-likely-avert-443
-million-skin-cancer-cases-united-states.

United States . . . The middle class woman of
the 1930s lived in a 'servantless household'—a
phrase you see repeatedly in scholarship about
this era. The refrigerator was tied to one of the
most fundamental and unifying of middle-class
events: the daily family meal. And it was in
providing for your family that the refrigerator
became a point of pride.*

In Britain, by contrast, things moved more slowly,
with just 2% of households owning the appliances by
1948. Rationing, which didn't end until 1954, was
undoubtedly a factor in the slow uptake in the British
market. By 1957, the year prime minister Harold Mac-
millan announced to the nation that "most of our people
have never had it so good," still only 13% owned a fridge,
rising to 58% by 1970. It would only be in the late 1990s
that Britain would catch up to the American level of
the 1950s.

The effects of the refrigerator, beginning in the
boom days of the West of the postwar era, are far-
reaching. Today, the vast majority of people in the
developed world have never lived in a home without
one. In the United States, thanks to that massive 96%
of households owning a fridge, this means that some

---

* "The Great Depression and the Rise of the Refrigerator," *Pacific
Standard*, October 9, 2012, updated June 14, 2017.

eight generations of people have never lived without an
icebox or a fridge.*

These people experience a world where fresh vegetables
and meat, all manner of dairy products, and ice are always
available. They come to take it for granted because they've
never known any different. In a refrigerated nation like
the postwar United States, the sheer quality of nutrition
available to a large population most of the time has far-
reaching consequences.

This is not something unique to the postwar developed
world. In 1999, as the new millennium approached, the
philosopher Umberto Eco was commissioned to write an
article about the greatest invention to shape the turn of
the first millennium, 1,000 years before.† Eco writes, "the

---

* The concept of a generation is somewhat nebulous. When you look
at a family photograph—say great-grandmother, grandmother,
mother, and child—it seems obvious: there are four of them
right there before you. However, no family has their offspring at
exactly the same time as another. Within the family photograph,
we can quantify a generation as being somewhere between 20–30
years. But, given the different times that different families have
children, this number then generally contracts. The so-called
Greatest Generation (1901–1927), those of age to fight in the
Second World War, is broadly deemed to last 26 years. The Silent
Generation (1928–1945) that follows between them and the Baby
Boomers (1946–1964) is allotted just 18 years by marketers, while
Generation Z (1997–2012) is assigned a mere 15. For the purposes
of saying "eight generations" above, many of the Lost Generation
(1883–1900) lived into the 1960s, from the height of the icebox
through Peak Fridge. And the generations that followed have never
been without these appliances.

† Umberto Eco, "How the Bean Saved Civilisation," the *New York
Times*, April 18, 1999.

Middle Ages before 1000 A.D. were a period of indigence, hunger, insecurity," before continuing,* "According to some scholars, Europe in the seventh century had shrunk to roughly 14 million inhabitants . . . Underpopulation combined with undercultivated land left nearly everyone undernourished." And yet, as the millennium turned, the population had begun to rise. In fact, by 1500, Europe's population had doubled, perhaps tripled.

The reason for this, Eco argues, is that sometime around the tenth century, Europeans worked out how to grow beans. And peas, and lentils. And these pulses saved their society. "When, in the tenth century, the cultivation of legumes began to spread," he writes, "it had a profound effect on Europe. Working people were able to eat more protein; as a result, they became more robust, lived longer, created more children and repopulated a continent."

His article is Euro– (and therefore, by the dubious virtue of colonization,† North America–) centric. By his own admission, Eco writes that he does not know enough about the history of legume agriculture on other continents. But his point is clear: without this surge in a better quality of nutrition, European societies could never have become as populous and economically successful as they did.

---

* A couple of paragraphs later.

† In short, without the population explosion in Europe, there would not have been enough white people to sail to the Americas and subjugate them.

In the twentieth century, as the second millennium approaches, the fridge takes on the role played by the bean a thousand years before. It injects nutrition into the population, supercharging whole societies.

The bottom line is that fridges make more, and stronger, people. So much so that there's an argument to make suggesting that the refrigerator bears significant responsibility for the Baby Boom.

There is a stubborn notion that the Baby Boom was caused by soldiers returning to their wives from the war. It has a certain romance to it—all these millions of Ulysseses returning lovingly to their Penelopes after so many years away. But, in fact, when you look at the numbers, as the economist Richard Easterlin did more than thirty years ago,* you see that the fertility of families in the United States whose fathers did not serve in the war tracks the fertility of veterans' families almost exactly. So, if the postwar Baby Boom was not caused by lusty homebound service men, it must be the result of something else.

In 1965, the Nobel Prize winning economist Gary Becker argued that

> the successful production of kids is subject to technological progress, just like other goods.

---

* Richard A. Easterlin, *Population, Labor Force, and Long Swings in Economic Growth: The American Experience*, Columbia University Press, 1968; and Richard A. Easterlin, *Birth and Fortune: The Impact of Numbers on Personal Welfare*, Basic Books: New York, 1980.

It will be argued that technological advance in
the household sector, due to the introduction
of electricity and the development of associated household products such as appliances
and frozen foods, reduced the need for labor
in the child-rearing process. This lowered the
cost of having children and should have caused
an increase in fertility, other things equal. This
led to the baby boom.[*]

Becker's suggestion here is for a hypothesis called
"appliance fertility," whereby access to laborsaving devices
frees up more time within a family not just for making
babies but for nurturing them as well.

We have to be careful about applying a direct causal
linkage between the proliferation of fridges and the Baby
Boom because, while the two things happen concurrently,
the act of buying a fridge does not necessarily, in the words
of Austin Powers, make you horny. But his is the whole
point of a hypothesis: to create a plausible theory that
we then attempt to disprove. What is not in doubt is this:
the invention of the fridge and its wider applications,
notably the "cool chain" (which is the subject of the next
chapter), fundamentally changed the way we nourish ourselves and indulge ourselves with food. This relationship

---

[*]   Gary S. Becker, "A Theory of the Allocation of Time." *Economic
       Journal* 75, 299: 493–517, 1965.

between improved human nourishment and population numbers is something we have seen before.

As Kaarin Goodburn, Secretary General of the Chilled Food Association (CFA), has suggested, the across-the-board improvement in food quality and safety offered by refrigeration is a significant factor in the increase in human life expectancy at birth that we see in the data from the advent of the fridge in the late 1920s on.* Access to better food increases the odds of both a fetus successfully coming to term and of the mother surviving the pregnancy.† It also dramatically reduces the odds of infant mortality. Thus the refrigerator,‡ like the legume one thousand years before, allows the human population to explode.

Meanwhile, the advent of the contraceptive pill, which became readily available in 1960, allows for reproductive choice. It is no coincidence that the Baby Boom is considered to end shortly thereafter in 1964.

More people means more pressure upon and more consumption of the resources available to us as a species. With more of us knocking about the place, and with the concomitant need for habitable space, it is the fridge's first

---

* Rick Pendrous, "Fridges Heralded the UK's Chilled Food Chain," *Food Manufacture*, August 24, 2017.

† Bearing children is, historically, one of the most dangerous things a human being can do.

‡ Along with other factors, most notably improvements in medicine, which also requires fridges, as we will explore in chapter 6.

cousin (air conditioning) that allows us to move to places we had never lived in vast numbers ever before.

An air conditioner works in almost exactly the same way as the fridge. It has a compressor, an expansion valve, and heat exchangers, and it pulls warmer air from the room it's there to cool, chilling it down and returning it just as if the room was your refrigerator cabinet. The hot air created in the process is then ejected outside.

And it was invented somewhat by accident.

In the summer of 1902, a young engineer, recently graduated from Cornell University, was assigned to solve a problem at the Sackett-Wilhelms Lithographing & Publishing Company in Brooklyn. His name was Willis Carrier. Their problem was that, in the summer heat and humidity, their ink couldn't dry fast enough and kept smudging in their printers. Carrier's solution was to build a refrigeration system that would chill the air around the print machines, condensing the humidity back into water. It worked. So well, in fact, that Carrier noticed that the print plant's workers wanted to enjoy their lunch breaks in the cool air next to the machines. He realized that his device had an application far beyond ink smudges: by removing heat from stifling summer rooms, he could make them not just bearable to enter, but actually pleasant. By 1906, after four years' work refining his ideas, he filed his first patent for an "Apparatus for the Treatment of Air."

While his early work focused on his invention's industrial applications—and the company he created to do so,

Carrier, continues to this day—he remained convinced that, as Steven Johnson writes, "air conditioning should belong to the masses."* In 1925, he persuaded Adolph Zukor, the driving force behind Paramount Pictures, not only that air conditioning would be good for his movie theaters, but to let him install an experimental system at the Rivoli, the studio's latest state-of-the-art cinema in Manhattan, as a demonstration. Zukor was as shrewd as you'd expect one of Hollywood's founding fathers to be. If there was to be a solution to the usual drop-off in box office takings over the summer,† he wanted to know about it.

Carrier's demonstration over the Memorial Day weekend, when Paramount was opening its latest Thomas Meighan‡ picture *Old Home Week*,§ was not without its hiccups.

---

* Steven Johnson, *How We Got to Now*, page 66.

† In a fairly typical story for its first issue after the Memorial Day Weekend, the trade paper *Variety* headlined a story in its May 27, 1925, issue: "AVERAGE DROP IN SUMMER FOR PHILLY," leading off, "Most picture houses have dropped to their normal summer pace. According to the usual figure, it is about $4,000 to $5,000 off in the bigger houses." This phenomenon would be seen at the box office across the United States.

‡ Meighan was a silent movie star who's largely forgotten today. But in the 1920s, he was commanding a solid $5,000 a week salary— about $84,650 today. Among the stars of the era whose names we still recall, he worked with Gloria Swanson, Anna May Wong, and Louise Brooks. Though he didn't know him, he was among those who bailed Rudolph Valentino out when he was arrested for bigamy in 1923, purely because he happened to be with one of Valentino's friends when the request for help came through.

§ *Variety*, May 27, 1925, and *Billboard*, May 23, 1925.

Among other things, to pass a safety inspection, he was asked to drop a lit match into his refrigerant. They were late getting started and the audience, already seated, was cooling itself with hand fans.

Carrier would later write of the event:

> It takes time to pull down the temperature in a quickly filled theater on a hot day, and still longer for a packed house. Gradually, almost imperceptibly, the fans dropped into laps as the effects of the air conditioning became evident. Only a few chronic fanners persisted, but soon they, too, ceased fanning. We had stopped them 'cold' and breathed a great sigh of relief. We then went into the lobby and waited for Mr. Zukor to come downstairs. When he saw us, he did not wait for us to ask his opinion. He said tersely, 'Yes, the people are going to like it.'[*]

They liked it so much that it became almost impossible to imagine a trip to the movies, unless at an outdoor screening or a drive-in, without climate control, even though it would take another fifty years for the summer blockbuster, as we understand it today, to catch on.[†]

---

[*]   Margaret Ingels, *Willis Carrier: Father of Air Conditioning*, 1952.

[†]   *Jaws*, released on June 21, 1975.

Throughout the 1920s and '30s, air conditioning in America would remain primarily in public commercial spaces like department stores, some of the more prestigious hotels and offices, and, of course, the picture houses. These systems were simply too large to be used in the home—though that didn't stop President Hoover from installing air conditioning in the Oval Office in 1929. It wouldn't be until 1931 that the single-room, window-ledge mounted air conditioner would be invented.[*] Priced at around $300,[†] these units were well beyond the means of the average consumer. It wouldn't be until the late 1940s that their sales began to take off.

By 2019, according to the US Census Bureau, a staggering 91% of American homes had central air conditioning, using some 6% of the country's electrical output to cool them down at a cost of roughly $29 billion a year. These are massive numbers. But lurking within that Census Bureau report is this: of the fifteen largest metro areas in the United States, only 44% of Seattle homes have air-con. This is revealing, because it reminds us that, in the temperate climate of the Pacific Northwest, air conditioning, while pleasant, is not a necessity. In southern US cities, like Miami or Houston, and in thousands of cities in the world's subtropical and tropical regions, it most definitely is. Air conditioning provides not only cooling

---

[*]   By H. H. Schultz and J. Q. Sherman.

[†]   Over $5,500 today.

comfort. It also changes our possibilities for living, both architecturally and geographically.

Consider the modern skyscraper: the Burj Khalifa in Dubai for example, its concrete, glass, and metal reaching for the clouds. It and so many other buildings like it would be impossible to imagine without air conditioning. Its interior would become so hot under the desert sun that it would be unbearable—a problem encountered in several skyscrapers built prior to modern air-conditioning solutions.[*] Many people assume that the windows on skyscrapers like this don't open to prevent accidents and suicides. In fact, it's to ensure the structure's climate control is not disrupted by the outside environment. Meanwhile, the building's air conditioning stops it from effectively turning into a greenhouse. Without this air conditioning, many of the structures we now take for granted could not exist. There would be no enclosed shopping malls or movie multiplexes. It would be impossible for the Middle Eastern emirate of Qatar to host football's 2022 World Cup, for which they purpose-built eight air-conditioned stadia.

---

[*]   The United Nations building in New York is a classic case in point. Built in 1948, it was designed by a board of architects, led by Wallace Harrison, that included, among other architectural luminaries, Le Corbusier. Le Corbusier advocating the use of *brise-soleil*—sun breakers—to shade the glass structure and interiors. He was overruled. Even though its windows are openable, and despite the installation of 4,000 Carrier air conditioning units, the west side of the building, in particular, roasted. Today, it costs almost $10,000,000 annually to keep it cool.

In cities like Houston, where underground air-conditioned concourses connect air-conditioned buildings with air-conditioned parking garages filled with air-conditioned cars, you can spend the vast majority of your day downtown without even venturing outside. Houston exemplifies another, arguably more profound, consequence of Willis Carrier's invention. It is a far from likely location for a major city. It is hot and humid. It is not brilliantly supplied with water. The city's British Consulate pays staff there a hardship allowance on account of the climate. Yet it has grown into the fourth most populated city in the United States.

This is down to a number of factors that could never have been anticipated when it was founded in 1837. The first was the dredging of the shipping channel, begun in 1915, that connects the Port of Houston to the Gulf, which allowed the city to become a vital trade hub. The second was the discovery of oil, which further turned it into an economic center. Add to that the massive demand for gasoline, rubber, and shipbuilding during the Second World War, and the 1961 decision to place NASA's Manned Spacecraft Center there, and you have all the conditions needed to create a boomtown. Between 1920 and 1950, the city's population would double twice to just under 600,000 people. The 2020 US Census now puts that number at over 2.3 million. Air conditioning, which makes living in such a climate bearable, plays a significant part in its growth.

In the years after the American Civil War, the general trend of population movement was from south to north. By 1964, that had reversed. In that decade alone, Houston's population grew to 940,000. Tucson, Arizona went from 45,000 to 210,000. The state of Florida, with a population of less than a million in the 1920s, had grown into the fourth most populous in the entire US by the 1970s. In the second half of the twentieth century, southern states from Florida to California, which had previously contained just 28% of Americans, became home to more than 40%. It is a migration made largely possible by air conditioning. And this massive shift in people from one end of the country to the other has fundamentally changed the nation's politics.

As Steven Johnson notes, "Swelling populations in Florida, Texas and Southern California shifted the electoral college toward the Sun Belt, with warmer climate states gaining 29 electoral college votes between 1940 and 1980, while the colder states of the Northeast and Rust Belt lost 31."[*] More than electoral college votes, this migration of Americans from north to south also shifted the political leanings of these sunshine states. Citizens heading south were not necessarily driven by job opportunities; many of them were looking to retire in the sun.

In his book *How Congress Evolves*, Nelson W. Polsby demonstrates how this demographic shift changed the

---

[*]   Steven Johnson, *How We Got to Now*, pages 69–70.

composition of the House of Representatives and its perfor-mance as a legislative body from the 1970s into the 1990s, initially liberalizing it as moderate Republicans took the place of anti–New Deal and anti–civil rights Dixiecrats before, in reaction to that change, coalescing into the par-tisan, ideological divides we see today. In fact, it has been argued* that this shift southwards of conservative Repub-lican voters was a key, if understudied, factor in the election of Ronald Reagan in 1980, at a time when America used more than half of the world's available air conditioning.

These air-con driven shifts in population are far from unique. In the rapidly developing city of Bangkok, the population has multiplied by a factor of eight in the seventy years between 1950 and 2020. In Mumbai, it has multi-plied by more than six and a half in the same period. So has Singapore.† The African cities of Nairobi in Kenya and Lagos in Nigeria have multiplied by a massive thirty-four times and forty-three times respectively.‡

Whereas the fridge, and the vastly improved nutri-tion it allows us to access, helps us make more people, air

---

* Notably by Steven Johnson in interviews given when he was promoting *How We Got to Now* in 2014, https://theworld.org/stories/2014 -10-04/how-air-conditioning-got-ronald-reagan-elected-president.

† Singapore's first prime minister after it gained its independence, Lee Kuan Yew, once referred to air conditioning as "one of the signal inventions of history," so radically has it allowed his country to transform.

‡ These numbers have been drawn from the United Nations' World Populations Prospects report of 2019.

conditioning moves us around the world so we can populate places where many of us would never have contemplated living before. This affects both our politics and the burden we place upon the planet's climate and resources. In the United States alone, air conditioning adds half a billion metric tons of carbon dioxide to the atmosphere each year.[*]

And yet, it is almost impossible to conceive the modern world without it. The internet, for example, could not exist without serious cooling technology. A big data center, housing the servers that allow us to access our files and photos from the cloud, uses the same amount of electricity as a town of 80,000 people.[†] Each one requires serious cooling.

Not all of that cooling comes from air-con.[‡] Since these servers do not have the cooling fans we see on home computers, and since their hard disks are hermetically sealed, they can be placed in cooling baths or, in Google's case, wastewater.[§] Facebook's European data center, in northern Sweden, is hydroelectrically powered and cooled by sub-Arctic air.[¶]

Meanwhile, skyscraper architects are falling back upon some of the ventilation solutions the pioneers of their

---

[*]   We'll be exploring aspects of climate change in chapter 8.

[†]   Tom Jackson, *Chilled*, page 234.

[‡]   Though quite a bit of it does.

[§]   Tom Jackson, *Chilled*, page 234.

[¶]   Ibid.

trade employed in the late nineteenth and early twentieth centuries. More excitingly, some are experimenting
with a concept called bioclimatic design. Writing for the
United States Green Building Council's website in 2014,
Alison Gregor reported that, "One example is a case
of biomimicry in Harare, Zimbabwe, where a mid-rise
building without air-conditioning was designed to stay
cool with a termite-inspired ventilation system. Scientists
digitally scanned termite mounds to map their architecture
in three dimensions, and then architects and engineers
applied the acquired knowledge about tunnels and air
conduits to create a blueprint for self-regulating buildings
for humans."* The firm T3 Architecture Asia has been
pioneering more green construction on projects such as
diverse affordable apartment blocks in Ho Chi Min City,
Vietnam, and luxury bioclimatic hotels in Cambodia.†

"The brute application of air conditioning," as the
American Institute of Architects described it in 1973, has
allowed construction companies to build structures of similar basic design almost anywhere in the world regardless of
local environmental constraints, be they homes, or malls,
or office blocks. This becomes a problem because, while
air conditioning units have shrunk over the years, their
basic workings have not radically changed since Willis

---

*   https://www.usgbc.org/articles/bioclimatic-design.

†   https://edition.cnn.com/style/article/t3-architecture-asia-bio
     climatic-architecture/index.html.

Carrier's day. While there have been improvements, they remain power-guzzlingly inefficient. Somehow, they seem to escape the notice of most climate legislators. As Stephen Buranyi wrote in 2019,* dealing with air-con efficiency falls under the "unglamorous label of consumer standards." Moreover, "Currently, the average air conditioner on the market is about half as efficient as the best available unit. Closing that gap even a little bit would take a big chunk out of future emissions."†

It comes down to the fundamental question that faces us as we seek to address climate change: If technology cannot find the solutions we need to bring down emissions and meet the ecological targets we set ourselves to do so, how much of our modernity are we prepared to surrender to safeguard humanity's future on this planet?

---

* "The Air Conditioning Trap: How Cold Air Is Heating the World," *Guardian*, August 29, 2019.

† Ibid.

5

Cool on the Move

Imagine for a moment that you're a bean, a green bean. A *haricot vert*. You're in Kenya, minding your own business, growing happily on your stalk in a region called Mwea on the eastern slopes of Mount Kenya, succored by sunshine and irrigation until you're ripe and ready. In less than forty-eight hours, you'll be flown in the hold of a commercial airliner to be sold in a British supermarket. A glorious, perfectly grown, out-of-season vegetable that UK consumers will happily shove in their shopping baskets without even thinking about how you got there.

In the food business, people will spend hours talking about provenance: how an ingredient was grown, where it was grown, how it got from producer to kitchen. The consumer, not so much; the consumer wants what they want when they want it, which, in the northern hemisphere, often means strawberries in January.

This is not normal.

In ordinary circumstances, we wouldn't be able to enjoy crispy fresh salads or raspberries in the winter. That we do is just another miracle brought to us by the cold: the cool chain, the way in which we move perishables, be they food or medicines, safely around the world.

These are staggeringly complex systems, with checks and paperwork at every step to ensure our Kenyan green bean, or indeed our Vietnamese frozen prawns, travel from farm to consumer, arriving unspoiled and safe to eat at the other end. They are both surprisingly robust, allowing millions of tons of food to travel the globe each day, and alarmingly fragile. Were the system to break down, even for a little while, the country affected would be in big trouble.

The United Kingdom currently has 71% of its land mass dedicated to agriculture,* which produces up to 60% of the nation's food needs. The rest must be imported; as of 2020, that meant 46% of it, much of it coming from Europe. Small wonder, then, that the UK's internal security service, MI5, works on the maxim that British society is never more than four meals away from anarchy.† Imagined scenarios include computer hackers throwing the food and

---

* United Kingdom Food Security Report 2021: Theme 2: UK Food Supply Sources, updated December 22, 2021, https://www.gov.uk /government/statistics/united-kingdom-food-security-report -2021/united-kingdom-food-security-report-2021-theme-2-uk -food-supply-sources.

† Will Iredale and Jack Grimston, "Britain Four Meals Away From Anarchy," *Sunday Times*, October 10, 2004.

water networks or the national grid into chaos, or bomb, biological, or chemical attacks preventing distribution.

MI5 maintains that, should something like this happen, it is more likely to occur in specific areas instead of nationwide. But even if that is the case, we cannot rule out the effects that rumors would have on the situation. At the start of the 2020 Covid pandemic, Britain somehow convinced itself that it was running out of toilet paper—it wasn't—and before too long, you couldn't find a roll of it in the shops for love nor money.

What MI5 did not imagine, however, was that the nation's food supply could be shut down by political self-harm. When Britain voted to leave the European Union in 2016, I doubt a single voter at the ballot box considered for a second that this might be a possibility. And yet, just two years later, the recently appointed Brexit secretary, Dominic Raab, had to announce to a parliamentary select committee that his department was now working to secure "adequate food supplies" in the event of a no-deal Brexit.*

This was, and is, because the food industry doesn't warehouse ingredients and products. It works on a "just in time" supply system, with most British supermarkets ordering the stock they need and no more. Most of it is stored within the cool chain itself, which makes the entire system vulnerable to the public mood. One whiff of rumor

---

* Jay Rayner, "Brexit provides the perfect ingredients for a national food crisis," *Guardian*, July 29, 2018.

or panic, as we saw in the pandemic, leads directly to mass buying of whichever product or foodstuff is alleged to be scarce, and thence to empty shelves, the sight of which simply enhances the fear.

Despite these fragilities, cool chains have transformed our world and our nourishment. The first road-going truck refrigeration system was invented in 1938 by a mixed-race American inventor named Frederick Jones. Jones was a natural mechanic and almost entirely self-taught. He seems to have been able to turn his hand to almost anything, building the transmitter for the first radio station in Hallock, Minnesota, inventing a machine to match sound with picture for the movies (which was later sold to RCA), and creating a ticket-dispensing machine for movie theaters.

Jones's truck refrigerator went through several iterations. The Model A worked, but was deemed too heavy to use. The Model B, though smaller and lighter, was not tough enough for the job. With the Model C, he cracked it. Mounted on the front of a lorry, it was light, compact, and durable, able to withstand the vibrations of the road.

Using the money made from the movie sound system's sale to RCA, Jones's boss, Joseph Numero, went into partnership with him, forming the U.S. Thermo Control Company, which would later evolve into the Thermo King Corporation, which remains one of the major US manufacturers of refrigerated vehicles. These machines allow the cool chain, as we understand it today, to exist.

They also make possible the very idea of frozen food, an idea that came about some twenty-two years before Jones's invention, in the midst of the First World War.

In 1916, a young naturalist called Clarence Birdseye decided it might be a good idea to move his young family north, to the tundra of Canada's Labrador province. Even back then, this was not a fun part of the world. Its climate is brutal. The food available left a lot to be desired. As Steven Johnson writes in his book *How We Got to Now*, "A typical meal would be what locals called 'Brewis': salted cod and hard tack, which is rock solid bread, boiled up and garnished with 'scrunchions,' which were small, fried chunks of salted pork fat."* Not something you're likely to find on the menu at a fancy New York restaurant any time soon.

However, as befits someone who would turn out to be a food innovator in later years, Birdseye was an adventurous eater. It wasn't long after his move that he began to go ice fishing with local Inuits. On these expeditions, he and his companions would cut holes in the ice of frozen lakes and cast for trout. It would be so cold that any fish pulled out of the water would freeze in a matter of seconds. What Clarence Birdseye would discover was this: when said fish was thawed out and cooked for his family, it tasted fresher than if it had traveled straight from lake to pan to plate. He was determined to find out why.

---

* Steven Johnson, *How We Got to Now*, Penguin Books, 2014.

His first thought was that the flavor of his trout was so good because it had frozen so close to its catching. But the more he looked into it, the more he discovered other factors were at work. Using a microscope, he discovered that food frozen more slowly had larger ice crystals trapped within it, adversely affecting its flavor. The quicker you could freeze a food, the better its preservation.

Today we call it "flash freezing."

Birdseye took a while to understand this and to turn it into a business. His was a gradual realization that frozen food could be a business. And it was only when he returned to New York and took a job with the Fisheries Association that the idea began to coalesce. Here he saw firsthand the awful waste that fishing boats faced with their catches without serious refrigeration. He would later write, "The inefficiency and lack of sanitation in the distribution of whole fresh fish so disgusted me that I set out to develop a method that would permit the removal of inedible waste from perishable foods at production points, packaging them in compact and convenient containers, and distributing them to the housewife with their intrinsic freshness intact."[*]

By the early 1920s, Birdseye had developed a process, modeled on Henry Ford's Model T car production process, to flash freeze fish at -40°C[†] on a double-belt freezer line.

---

[*] Steven Johnson, *How We Got to Now*, Penguin Books, 2014.

[†] The temperature where Fahrenheit and Celsius meet . . .

He found that it worked for other ingredients, too: fruit, meat, vegetables.

But, at this point, Birdseye was ahead of the curve. America may have been embracing the fridge and the freezer, but not yet in enough numbers to make his vision viable. By 1929, with the crash that created the Great Depression looming (though no one knew it), it was. Thanks to Clarence Birdseye, restyled "Captain" for marketing purposes, we can eat fresh-frozen peas and cod, and a host of other foods, whenever we want, seasonality notwithstanding. And regardless of the green beans flown in fresh from Kenya.

It remains an interesting question: fresh or frozen? When it comes to peas, in my personal experience, frozen trumps fresh bought at the supermarket. But fresh picked from plants I've grown myself wins every time. It's almost like you can taste the earth in them. As a consumer, I've been more than happy with a frozen pea or broad bean (or fava bean), while I'm not wild about a frozen *haricot*. I don't feel it has the integrity of its fresh counterpart. And this brings up an interesting question: that of the morality of buying a bean that comes all the way from Kenya.

On the one hand, by doing so, I'm helping to create jobs in developing countries. As it stands, the Kenyan green bean market is dominated by small farmers. There are, according to recent reports, about 50,000 smallholders working on farms of less than a hectare producing huge amounts—an average of about 19,000 tons a year—of vegetables for the British market. While these exports are ferried in the holds of passenger planes shipping European tourists back from safaris and beach holidays, they pump out a lot of carbon emissions, affecting our environment.

It is 8,640 km from Nairobi to London. Yet these imports account for just 5% of Britain's total aviation carbon emissions. Tourism, curiously, accounts for 90%. And yet, while human travel accounts for the vast majority of emissions, it's arguable that our food travels further than we do. It doesn't matter where we are in the world: there is something we consume regularly that doesn't grow where we live. Coffee, chocolate, mangoes. Strawberries in winter.

Barring political self-harm and our carbon emission making choices, we can enjoy them all.

6

Your Cold, Cold Heart

When we read fiction, the celebrated semiologist and critic Umberto Eco tells us,[*] we experience three kinds of time: story time, the amount of time that elapses in the course of the tale—you can write, "100 years passed" and, though it takes but seconds to read, in the story those years have indeed passed; discourse time, the time the story takes to tell; and reading time, the days or weeks it takes us to get through a book. When we watch a movie or a TV show or a play, we experience just two: story time and discourse time, the latter defined by the length of the production.

I mention this because, while the discourse time of Quentin Tarantino's *Reservoir Dogs* is a fleeting one hour and thirty-nine minutes, the story time, flashbacks excluded, is basically how long it takes for Tim Roth's character, Mr. Orange, to bleed out.

---

[*] Umberto Eco, *Six Walks in The Fictional Woods*, Harvard University Press, 1994.

Gut wounds like that are, by all accounts, a horribly painful and slow way to die. But many other shootings and stabbings cause rapid and massive blood loss. Should you be the victim of such an awful thing, your odds are not good, despite everything you've seen at the pictures—because the movies are all about heroes. The screenwriter William Goldman* once wrote about a conversation he'd had with a veteran New York firefighter.† He asked him who was the best firefighter he ever saw, the Willie Mays of firefighters, if you will. The veteran said, well, there was this one guy, and proceeded to tell a story of a man who, despite all the noise and heat and collapse of the fire they were fighting, heard a baby, alive in the building, that no one else could, and then successfully saved it. Good story, wrote Goldman. But the problem for the screenwriter, he said (and it's true), is that that's what Tom Cruise does in the first reel.

Then you have to top it.

The same goes for the doctors in our dramas. Whether the man, woman, or child, shot or stabbed, wheeled in on a gurney, will or won't be saved is based not on their medical situation but on where we find ourselves within one doctor or nurse's character arc. Or, to put it another way, is it time in the show for the hero to suffer, like Dr. Greene in *ER*'s horrific pre-eclampsia storyline, or

---

*   One of my heroes.

†   William Goldman, *Which Lie Did I Tell*, Pantheon Books, 2000.

like Dr. Ross, in the same show, to heroically rescue a drowned child from a storm drain?

Sad to say, it just doesn't work like that in a real-life trauma unit. Should you lose half your blood volume, about two and a half liters, as in these kinds of situations you are likely to do, you have a 5% chance of survival. As the trauma surgeon and researcher Samuel Tisherman told *WIRED* in 2020, "Surgically, it's a race against time to get the bleeding stopped so you can resuscitate the person before their internal organs are damaged irreversibly by not having enough blood flow."[*]

Tisherman thinks he may have a solution to this problem: the discrepancy between the time it will take to heal you and the time you have left. It lies in an interesting place. At the beginning of this book, we explored step by step how the cold kills. In this chapter, thanks to people like Tisherman and others, we will look at how, in the right hands, it can also save your life.

In the words of Dr. Mads Gilbert, who successfully revived the Swedish radiologist Anna Bågenholm after she'd fallen through ice when she accidentally skied over a frozen stream in 1999,[†] "Nobody is dead until they're warm and dead." Which begs the question: Can we chill

---

[*] Nathalie Healey, "Extreme cold is bringing humans back from the brink of death," *WIRED*, February 5, 2020.

[†] Ibid.

someone down until they are effectively dead, do the work required to save them, and then warm them back to life?

Within the answer—which is "yes," by the way—lies a number of fascinating tales, many of them wrapped up in one single concept: medical hypothermia. Or, as it is more properly called, emergency preservation and resuscitation, or EPR.

As radical as this seems, it is, surprisingly, not a new idea. In fact, it's so old that the first person to recommend using ice or snow in medicine was Hippocrates, back in the fourth century B.C.E.—his recommendation being to pack soldiers' wounds with ice or snow, if available, to slow the bleeding. In 1812, Napoleon's surgeon, Dominique Larrey, used ice on soldiers to provide pain relief during amputations. While these were both highly practical applications of the cold in a medical setting, they were miles removed from the ways in which doctors have induced hypothermia for at least seventy years.

Its first advocate was the pioneering Canadian surgeon Wilfred Bigelow. According to his obituary in the *Lancet*,* he first began thinking about the idea while he was working as a surgical resident in Toronto, where he often found himself having to amputate patients' fingers or toes damaged by frostbite. Then, later, when he served in the Canadian Army Medical Corps during the Second World War, while treating similar cases of frostbite, he noticed not only that

---

*   May 7, 2005.

the cold reduced both the metabolic rate and oxygen needs of the affected area, but that if he could only warm them up more slowly, he could reduce the damage to the tissues. He even had a machine constructed to make this possible.

It wasn't until after he'd returned from the war and started to train in cardiovascular surgery at Johns Hopkins University that he would realize cold's true potential in the medical world.

Once upon a time, it was generally believed that heart surgery, in any form, was impossible for one blindingly obvious reason: if you stop the heart moving so you can make an incision into it, the patient's brain will be starved of oxygen and they'll die. This didn't stop surgeons from trying. As early as 1801, a Spanish doctor called Francisco Romero successfully cut a fistula, a window into the outer sac that surrounds the heart, in a procedure now known as a pericardiostomy, to drain fluid trapped between the inner and outer linings. According to his memoir, he pulled this off five times, with only one death. A similar surgery would be successfully performed in North America ninety years later, when Henry Dalton, a surgeon from St. Louis, Missouri, successfully sutured the pericardium on September 6, 1891, a procedure that would not be repeated on the continent until Daniel Hale Williams, the pioneering African American surgeon, was able to replicate it on July 10, 1893.

Williams was a trailblazer in more ways than one. Among his many achievements, he founded Provident

Hospital in Chicago which, in addition to providing much needed health care to the city's African American community in days when it was still broadly segregated, was also a training facility for the nurses and doctors from the community. In 1895, he created the National Medical Association for African American Doctors and campaigned tirelessly for racial inclusion in US medicine in general. By 1913, he had become the only African American doctor in the American College of Surgeons.

Despite pioneers like these and the huge advances made in cardiac surgery over the next fifty years, the central problem still remained: the beating heart. For it is nigh on impossible to make an accurate incision into a moving object.

In 1950, Bigelow put together a team to explore his hypothesis: if you chilled down a body sufficiently, you would massively decrease its need for oxygen; therefore you could stop the heart for long enough to operate upon it without oxygen depletion causing brain damage in the patient. If it worked, open heart surgery as we understand it today would become possible.

Meanwhile, and unknown to them, an American doctor called John H. Gibbon was working on another solution to the same problem. In 1931, when he was a surgical research fellow at Harvard, Gibbon had taken part in an operation to save a patient with a severe pulmonary embolism. She didn't make it. But Gibbon was left with the idea that, if he could invent a machine to bypass the heart and lungs which would oxygenate the blood as it passed through, he could save the life of any other patient like her.

Gibbon, who performed his research throughout the 1930s and 40s, favored using cats as his test subjects. Bigelow, on the other hand, worked on dogs, operating on 176 of them in the course of his experiments, of which only 51% were successfully revived.[*] He faced a key problem, particularly in his early attempts: the unconscious animals would develop a form of arrhythmic heartbeat known as ventricular fibrillation. The other problem was that the dogs stopped breathing when their core temperature dropped lower than 24°C (75°F).[†]

Before Bigelow began his work, hypothermia was something to be avoided at all costs. But Bigelow was certain he was onto something. Of the many questions that they were exploring, including the best ways of cooling a patient and how long they might be kept under for, one stands out: Would it be possible somehow to recreate the conditions of hibernation—in which groundhogs, in particular, are able to survive at temperatures as low as 3°C (37.5°F)—in human beings?

---

[*]   Lara Rimmer, MBChB, Matthew Fok, MBChB, Mohamad Bashir, MD, MRCS, "The History of Deep Hypothermic Circulatory Arrest in Thoracic Aortic Surgery," *Aorta*, June 2014, vol. 2, issue 4: 129–134.

[†]   Ibid.

He started working with rhesus monkeys and the aforementioned groundhogs to find out. While he was able to artificially induce hibernation in the groundhogs, this wasn't possible with the primates. But he discovered that, if you cooled them down to 20°C (68°F), unlike the dogs, they kept breathing at a rate of just eight inhalations per minute.

Bigelow had one final conundrum to crack before his research could move to the next level: shivering.

As we've discussed previously, shivering's job is to warm us up. When it kicks in, it raises our metabolism and our heart rate fails to slow. But, as Bernard Goldman, professor of cardiac surgery at the University of Toronto, told the *Lancet* for its obituary on Bigelow: "His research on the dog was initially frustrating because they shivered when they got cold and their metabolism increased. Once he figured out how to control shivering, he saw there was a linear relationship between metabolism and temperature. He realized you could slow down the heart." Bigelow had used the cold to buy a fifteen-minute operating window on a stopped heart.

Parenthetically, it was also during these experiments that Bigelow stumbled onto the idea behind the pacemaker. During one of his early procedures, the dog on the operating table's heart stopped and would not respond to cardiac massage. "I had this probe in my hand," he would say later. "Out of interest, and in desperation, I gave the heart a poke. To my surprise, it immediately contracted."*

Bigelow presented his findings, on both hypothermia as a means to enable heart surgery and on the idea that would lead to the pacemaker, in 1950, first in Denver to the American Surgical Association, then in Boston to the American College of Surgeons. An American surgeon called F. John Lewis was present at the former and was fascinated by Bigelow's work. So much so that, on September 2, 1952, he employed the technique, still known at the time as medical hypothermia, to close an atrial septal defect in a five-year-old girl. It was the first successful performance of open-heart surgery, with the heart itself open directly to the surgeon's gaze, the world had ever seen. In Bigelow's words, Lewis had broken the ice.

While Lewis continued to put Bigelow's work into practice throughout the 1950s and press on with the research, it was, thanks to the competition of John Gibbon's bypass experiments, not to catch on. Gibbon deployed his new machine for the first time on May 6, 1953, bypassing his patient's heart and lungs to open up the heart and perform the same procedure Lewis had completed under very different conditions. The heart and lung bypass machine would go on to become the standard preference in cardiac

---

* Thomas H. Maugh II, "Wilfred G. Bigelow, 91; Cardiac Specialist Invented Pacemaker," *Los Angeles Times*, April 1, 2005.

surgery in North America and the West because medical hypothermia allows only for a relatively short operating window. As Samuel Tisherman noted (above), as in emergency medicine, time is often the enemy.

However, in the Soviet Union, and in Novosibirsk, Siberia, in particular, things were different. In 1998, a team of ex-Soviet surgeons traveled to the Medical Center at Penn State University to undertake a collaborative research project into the uses and practicalities of employing hypothermia in cardiac surgery. While the Americans had moved on from hypothermia, thanks to bypass machines, the Soviet Russians had embraced it, not simply because the machines could be hard to come by, but because there are distinct advantages to, well, *not* removing all the blood from a patient's body and putting it back in.

The issue is that, following open heart bypass surgery, a significant number of patients experience emotional and behavioral changes. On the one hand, this is probably to be expected. Having your ribcage cracked open is a traumatic experience, even if you're anesthetized. Even though most patients return to their old selves, usually after about six months,* doctors would come to realize that there is something else going on. In short, your blood leaves your body. It enters a foreign space—the machine. As a part of your immune system,

---

* Osama Younes, Reham Amer, Hosam Fawzy & Gamal Shama, "Psychiatric disturbances in patients undergoing open-heart surgery," *Middle East Current Psychiatry*, September 17, 2019.

blood is highly evolved to respond defensively to the presence of alien materials, like the viruses and bacteria that cause disease. Your antibodies don't recognize this space. Even though they have no idea what it is, they arm up.

The first place freshly oxygenated blood goes, once it leaves the heart, is the brain, your two most vital organs working in tandem to keep everything else in the evolutionary miracle of your body going. Except here, today, as your operation takes place, your blood isn't coming from your heart but from a machine. The antibodies it carries, now activated by the strange surroundings of the airtight metal machine, are primed to attack. Which is exactly what they do. As confused as they must be, their target quickly becomes the first place they visit next: your brain.* In addition to the resulting inflammation of the brain, there is also a risk of blood clots forming within the machine, which can cause aneurysms when they return to the body. The Penn State team wanted to explore whether medical hypothermia would offer a solution.

Back when Dr. Lewis performed his first open heart surgery in 1952, he was able to stop the organ for just fifteen minutes after reducing his patient's body

---

*   Even though I've known about this concept for a while, when my father-in-law had his (successful) open heart surgery in 2006, I was shocked by the way it turned this kind and gentle man into an arsehole for about five days. He soon came back to himself.

temperature to 10° below normal.* The team from the Novosibirsk Institute of Circulatory Pathology, led by Dr. Vladimir Lomivorotov, now claimed that they could stop one for up to ninety minutes, greatly widening the surgical window. That they could do so was thanks to the work of pioneering Soviet doctors like Professor Yevgeny Meshalkin and Dr. Vasilii I. Kolesov in the 1960s.

Kolesov in particular invented a procedure known as coronary artery bypass grafting, or CABG, which has gone on to become one of the most successful and frequently performed operations in the world. Yet, from February 25, 1964 to May 9, 1967, Kolesov's surgical department in Leningrad was the only place on the planet that could do it.[†]

Kolesov was convinced that, while the heart bypass was both a safe and reliable method for performing cardiac surgery, it just wasn't worth the risks of the brain inflammation it caused. He was so convinced that, between 1964 and 1974, he only used one in just 18% of his CABG operations. But even though he made this decision out of preference, it is worth noting that Soviet hospitals of the era were spectacularly under-resourced, especially when compared to their American equivalents, a legacy of war. During the Siege of Leningrad alone—a horrific event

---

*   Richard A. DeWall, "The origins of open heart surgery at the University of Minnesota 1951 to 1956," *Reflections of the Pioneers*, vol. 142, issue 2, pages 267–269, August 1, 2011.

†   Igor E. Konstantinov, "Vasilii I. Kolesov: A Surgeon to Remember," *Texas Heart Institute Journal*, 31(4), pages 349–358, 2004.

Kolesov not only lived through but continued working as a surgeon during, despite huge privation*—the Soviet Union lost twice as many people as the entire US forces' casualties in the entire war.

At least Kolesov had access to bypass machines and a choice whether to use them or not. In Novosibirsk, the use of medical hypothermia to slow and stop the heart was the only option. Despite the technique's previous use in North America, Japan, and Great Britain—even the ground-breaking surgeon Christiaan Barnard gave it a go in 1963—by the 1990s the Novosibirsk Institute's surgeons were the only people in the world who still used it. While the Penn State/Novosibirsk Institute match up ultimately came to nought, it was not the end of the road for medical hypothermia because of the very thing that made it so useful for heart surgery: it's all about protecting the brain.

Today the procedure is known as deep hypothermic circulatory arrest, or DHCA, and it is, in effect, controlled clinical death. It works like this: you begin by placing your anesthetized patient onto a cardiopulmonary bypass system. As it takes over from the patient's heart and lungs, you then begin to remove and store a percentage of their

---

*   On one occasion, a German bomb exploded behind his hospital, blowing out the operating theatre's windows. A shell fragment missed his head by centimeters, while another thunked into the ceiling. The only thing Kolesov could do to protect the patient from debris was to shield the prone body with his own.

blood for later use, and replace it with other fluids for two key reasons: you want to prevent the blood from thickening and clotting as it cools, and you want to remove as much glucose from their system as possible to avoid developing hyperglycemia.

Now the cooling process begins within the bypass machine, chilling the diluted blood to lower your patient's temperature from within. This proves to be a more stable method to the external cooling practiced in Novosibirsk, which both lacked precision and can lead to skin irritations and shivering, which we really don't want to happen.

Your goal here is to bring your patient's core temperature down to between 20–25°C (68–77°F), to a point where the heart stops and the brain enters a state of electrocerebral silence. It's time to switch off the bypass pump. Your patient's circulation stops. You and your team now drain still more blood from their body, and now, in a bloodless surgical field, you have about an hour to fix whatever happened to their brain. When you're finished, you reverse the process.

This is the tricky part.

While the brain and heart will restart naturally, the latter can sometimes fibrillate, or beat arrhythmically, requiring your intervention to bring it back to normal. Most critically, as you rewarm your patient, you *cannot* overshoot a core temperature of 37°C (98.5°F). To do so risks undoing all your work thus far, disabling their brain to the extent that, in the worst-case scenario, you could surrender them to living the rest of their days in a vegetative state.

Be in no doubt that this is a risky procedure. Between 3–12% of patients will experience some form of neurological injury, despite your best efforts in the operating theater. Almost all of them will have issues with their glucose metabolism thereafter and require insulin injections to control it. However, not every brain cooling technique is so extreme. In fact, the one that is perhaps the most important requires no surgical intervention at all. It was developed to prevent brain damage in newborns.

The key problem here is something called birth asphyxia which, of the 750,000 births in the United Kingdom each year, affects around one thousand. When the brain is starved of sufficient oxygen, even in the first few moments of life, it produces harmful toxic chemicals. It continues to do so even after its oxygen supply has returned to normal levels. The resulting damage is often not detectable until between eighteen months to two years into the child's development.

As said before, childbirth is one of the most dangerous things a human being can go through. Even with the most outstanding medical care, awful things can happen. I'd suggest, if you're squeamish or likely to be triggered by such things, you might want to skip the next few paragraphs.*

---

* Writing them left me relieved that my wife and I never wanted to have children. If I'd known about this stuff when we considered it, it would've scared the bejeezus out of me.

The first potential cause of this is called nuchal cord, whereby the umbilical cord gets, by the fetus's movements within the womb, wrapped around its neck. Most of the time, this is entirely harmless, and the cord unwraps itself. But, every so often, it develops into something called tCAN syndrome which, believe it or not, stands for "tight cord around the neck." This can become a problem, for obvious reasons, and is fortunately quite rare.

The next issue is something called placental abruption, also rare, occurring in seven to twelve out of every thousand births in North America. Here the placenta comes away from the wall of the uterus prior to birth, depriving the baby of nutrients and oxygen.

Then we have uterine rupture. This occurs in one out of 1,146 pregnancies, and is more likely in people who've already had some form of uterine surgery before. Here the womb rips and the fetus, the placenta, or both enter the abdomen.

Umbilical cord prolapse can happen in one to six out of a thousand deliveries. Here, when the mother's waters break, the umbilical cord escapes through her partially dilated cervix. Should the cord then be squeezed between the bodies of both mother and fetus, the baby will be starved of blood and oxygen.

Finally, and rarest of all, is something called an amniotic fluid embolism, which occurs in one out of 40,000 deliveries in North America. Here, amniotic fluid enters the mother's blood stream, sending her into shock, and potentially cardiovascular collapse, serious bleeding, even the loss of the ability to breathe.

Each of these things are terrifying. They are, fortunately, (as I've said before, but I think it bears repeating) rare. Thanks to the cold, help is at hand in each situation.*

Research shows that, as soon as any of these situations occurs, doctors have up to six hours to take action to prevent permanent damage to the newborn's brain.

Sometimes, the baby is placed on a special cooling mattress. In other situations, a cooling cap will be placed upon its head, or it will be swaddled in a special cooling blanket. Whichever device doctors choose, the goal remains the same: to bring the body temperature down to 33.5°C (92.3°F), inducing a state of mild hypothermia for three days.

The results are astounding, so much so that this has been called the greatest breakthrough in neonatal care in forty years.[†] However, according to an article in *Nature*,[‡] studies on the technique's use in low-to-middle-income

---

* I have sourced the above from the UT Southwestern Medical Center's blog, on a post written by Robyn Horsager-Boehrer, MD, and Becky Ennis, MD, July 23, 2019.

† My source for this is anecdotal, and said by a nurse to a friend of mine whose granddaughter was treated, post birth, with the cooling cap.

‡ Max Kozlov, "Doubts raised about cooling treatment for oxygen-deprived newborns," *Nature*, August 11, 2021.

countries (LMICs) have found it not only not to work, but to show links to increased infant mortality.

The research, published in the *Lancet* on August 3, 2021, looked at a sample of 202 babies born with suspected brain damage in India, Sri Lanka, and Bangladesh. All received the three-day cooling treatment. 42% of them died within eighteen months. By contrast, 206 newborns with similar symptoms in the same countries received more traditional treatments; of them, only 31% lost their lives.

Consequently, the lead researcher, Sudhin Thayyil, a perinatal neuroscientist at Imperial College London, called for the technique not to be used in these regions until further research could take place.[*]

This is very bad news, because the majority of infant disabilities and deaths caused by the awful circumstances described above happen in LMICs. The reasons behind it may be as simple as the notion that richer countries perhaps have access to better health care than their poorer neighbors. Quoted in the same article, Joanne Davidson, from the University of Aukland in New Zealand, said, "These babies are not presenting with the same kind of patterns of injury as in a first-world setting." Perhaps, as the saying goes, different diseases require different remedies.

---

[*]   Max Kozlov, "Doubts raised about cooling treatment for oxygen-deprived newborns," *Nature*, August 11, 2021.

As we've observed before, these kinds of treatments, be they reviving a frozen patient who has fallen through ice to one placed in deep hypothermia for heart surgery to the infant only mildly cooled to protect their brain, are highly complex. It only takes one or even two things to go ever so slightly wrong for the whole procedure to fall apart. Presenting the contrary opinion in the same article, Marianne Thoresen, a neonatal neuroscientist at the University of Bristol, argues that part of the problem could be that, in LMICs, the nurses looking after these infants may be spread too thin while, in first world hospitals, the standard is to have one nurse for each baby undergoing such a treatment. In the article, she says, "These babies need very high-quality intensive care, because when you cool somebody, a lot of the physiology is affected."*

In the latest work on therapeutic hypothermia, very high-quality intensive care becomes very high-quality emergency care. That care is time critical. Which brings us back to Samuel Tisherman and his work to buy that time for trauma victims on the operating table. The issue he is trying to solve is one that has confronted surgeons since the invention of medicine.

The human animal contains approximately five-and-a-half liters of blood, which accounts for 8% of its body weight. Should you be shot or stabbed, or bitten by a

---

* Ibid.

shark, one of the very first tasks confronting paramedics or bystanders trying to save your life is to get the bleeding under control. Because, while you can stand to lose up to 30% of your blood volume (under such conditions, your breathing, heart rate, and blood pressure will remain close to normal), any more than that will lead to trouble. Your heart will start to beat more than 120 times a minute. Your breathing becomes more rapid. Your blood pressure falls. You, your paramedics, and your surgical team, should you make it to the emergency room, are now in a race against time to save your life. Because, if you lose more than 40% of your blood, you face irreversible brain damage in just five minutes, and permanent heart failure in twenty.

Tisherman's solution is, in his words, to induce a state of hypothermic preservation, chilling the patient rapidly to significantly reduce the internal organs' need for oxygen, buying the surgeon time to repair the bleed.

As of February 25, 2022, his extraordinary procedure, formally called Emergency Preservation and Resuscitation for Cardiac Arrest from Trauma,[*] has entered phase two of clinical trials, which means his team at the University of Maryland are actively looking to recruit an estimated twenty participants to test it.

Tisherman has written that, using the best of the current emergency surgical techniques, which include

---

[*]   He began working on it in October 2016.

a thoracotomy,* exsanguinating trauma patients who go into cardiac arrest have, at present, a less than 10% survival rate.† More recently, he wrote that, in Phase 1 clinical trials, working with large animals, he has demonstrated that, if you cool the tympanic membrane, which lies within the ear, to 10°C (50°F), you buy a window of up to two hours to make the necessary repairs to any trauma and allow a full neurological and cardiac recovery.‡

The technique involves inserting an arterial catheter into the descending thoracic aorta, and then injecting ice cold saline into the body to rapidly cool the brain to under 10°C (50°F). This buys the surgeon time to repair the damage before rewarming and resuscitating the patient on a full bypass. In effect, it places them into a state of suspended animation.

If this sounds like science fiction, brace yourself for the next bit. Recruitment for Phase 2 of clinical trials may have, at the time of writing, only just begun. But, on November 20, 2019, *New Scientist* magazine exclusively

---

* Which covers any operation wherein the surgical team need to get inside your ribcage.

† Samuel A. Tisherman, "Therapeutic Hypothermia in Severe Trauma," *ICU Management & Practice*, vol. 13, issue 4, Winter 2013/2014.

‡ Samuel A. Tisherman, "Emergency preservation and resuscitation for cardiac arrest from trauma," *Annals of The New York Academy of Sciences*, vol. 1509, issue 1, March 2022.

reported that Tisherman had already done it.* In the article, Tisherman himself is quoted as saying it was a little surreal when they tried it for the first time.

This is a radical treatment and, given its "race against time" nature, one where, in authorizing it, the US Food and Drug Administration exempted it from the usual need for patient consent, given that, without it, they'd certainly die.

So far, Tisherman has remained tight-lipped about their success rates, though we will find out when Phase 2 ends in December 2023. He seems cautious about the science fiction overtones of the phrase "suspended animation," telling *New Scientist*, "I want to make clear that we're not trying to send people off to Saturn. We're trying to buy ourselves more time to save lives."[†] But, as it stands, his team's work offers a literal lifeline to the doomed by taking a mechanism that kills us—hypothermia—and flipping it to our advantage.

Here, in the emergency room, cold is the new frontier.

---

* Helen Thomson, "Exclusive: Humans placed in suspended animation for the first time," *New Scientist*, November 20, 2019.

† Ibid.

7

Fun and (Ice) Games

Once upon a time, when I was ten years old,* we used to have a favorite winter game at school. When the weather was cold enough, we'd bring cups of hot water out on the playground before class and pour it in long strips on the tarmac. By break, which was at 10:45 in the morning, these would have frozen into ice runs. The idea was then to run full tilt at the head of this line of black ice and . . . slide.

Why the teachers never stopped us, I'll never know. But this game was indulged until somebody hurt themselves and all the kids were sent to play indoors until break was over.

In the winter of 1982, that somebody who hurt themselves was me.†

---

* Please excuse the personal anecdote here, I promise you it's going somewhere.

† I fell on my face and smashed my teeth out.

I mention this because, in one of those strange coincidences one finds when researching a book like this, 174 years and one week prior to my silly accident, pretty much to the hour, a Norwegian-Danish army officer called Olaf Rye launched himself, before an audience of fellow soldiers gathered outside the Eidsberg church in Østfold, Norway, a staggering nine-and-a-half meters through the air in the world's first recorded ski jump.

Captain Rye did not, on this occasion, knock out his teeth; though, in the course of a celebrated military career, who knows, he may have suffered some kind of dental accident. Instead, he had a square named after him in Oslo, among many other honors, and he remains a skiing pioneer and Danish national hero.

Despite Rye's outrageous feat, ski jumping would not become a competitive sport for another fifty-nine years, when the first event was staged in Hauglibakken[*] on March 8, 1868.[†] But it's appropriate that the act was both first performed and staged as a sport in Scandinavia. For it is in the lands of the Norse that skiing was most likely invented. Certainly, it is where much of our first evidence of the practice can be found.

Skiing, skating, and sledging have existed for as long as humanity has lived with snow and ice. But their

---

[*]   Also in Norway.

[†]   It was won by Sondre Norheim, another Norwegian who later emigrated to the United States, and another of skiing's pioneers.

permeation into the wider world in the form of recreation is a relatively recent phenomenon, and broadly
concurrent with the story of our adoption of ice and
refrigeration to preserve our food and ship it around the
world. This is, in no small part, connected to the same
wider social forces that have shaped our air-cooled and
refrigerated world.

Throughout the mid-to-late nineteenth and early
twentieth centuries, we see an explosion of recreation.
It ranges from the proliferation of pubs and bars to ice
cream parlors to the codification of sport as we understand it today. The first draft of the *Laws of the Game*, for
example, which laid out the rules of Association Football,
was written in 1863. The first indoor athletics meeting was
held that same year. Cricket, which is known to have
existed in one form or another since 1597, becomes the
game we know today in 1864, when over-arm bowling
was legalized and the first edition of *Wisden** appears. All
of this happens because, suddenly, more and more people
have time on their hands to indulge in leisure.

Previously, as with the use of ice, leisure was a luxury
the vast majority could not afford or even dream of. Its
democratization in this period bears examination through
the prism of winter sports and our messing about in the
snow. It changed humanity's relationship with some of our

---

* The so-called "Bible of Cricket," published annually in the United
Kingdom.

harsher environments and our species' growing footprint upon our world.*

Skiing itself dates back for thousands of years. But, for the vast majority of those years, it was not a recreational activity. It was a practicality.

As far as we can tell, it originated in Scandinavia, or China, or Russia, or Central Asia. The evidence is patchy at best: the oldest pieces of it include 5,000-year-old paintings found in the Altaic region of China; fragments of ski-like objects discovered at Lake Sindor, some 1,200 km northeast of Moscow; and the Rödöy petroglyph, also some 5,000 years old. Scratched into rock on the Norwegian island of Tro, it depicts a stick figure on skis. In Norway it is iconic. It inspired the logo for the Winter Olympics in Lillehammer, held in 1994, and was tragically damaged in 2016 when someone,† with apparent good intentions, rescratched over the image to try to make it more visible.

The ski in the image has been described as being too long and too upturned at the front to be practical. In his article, "The Ski: Its History and Historiography," the writer LeRoy J. Dresbeck suggests it either depicts a very early prototype, or that the image displays certain artistic

---

* Some of it's quite fun, too.

† A youth, according to news reports at the time, some of which said "youths." Norwegian proceedings have protected their anonymity.

liberties,* leaving the artist's proficiency at drawing or
scratching into rock uncommented upon. Nevertheless, there it is: a human on skis. Unique in the world.
Telling us purely that skiing existed, and with no context about the practice's invention or story before it was
drawn—such are the frustrations of prehistoric evidence,
which tells us always "that," never "how" or "why," and
offering only tantalizing glimpses of what existed and was
possible before the invention of writing could explain
things to us further.

For written evidence, we must venture further forward in time. Some say, wrongly, that the first of this
appears in Xenophon's *Anabasis*, an extraordinary auto-
biographical account of a campaign fought by an army
of Greek mercenaries hired by Cyrus the Younger to
seize the throne of the Persian Empire from his brother,
Artaxerxes II, in 401 B.C.E.

In his fourth book, Xenophon describes snowshoes,
worn by horses, in the mountains of Armenia. Even
though there is evidence of skiing in Central Asia, and
some have speculated† that this is where the practice truly
began, this is not it. For the real deal, we have to wait
another 800 years, until the sixth century C.E.

---

* LeRoy J. Dresbeck, "The Ski: Its History and Historiography,"
  Technology and Culture, vol. 8, no. 4 (October 1967), pages
  467–479.

† And been debunked.

In both Jordanes's *Getica** and Procopius of Caesarea's *De Bello Gothico*, we find references to peoples of the north, most likely Scandinavian Finns, skating or gliding on special equipment to cross the snow. While Procopius is often considered to be the last of the great historians of the ancient western world,[†] Jordanes is a trickier source. The *Getica*, Jordanes tells us, is essentially an epitome of a lost history of the Goths by Cassiodorus, an Italian who served as the *magister officiorum*[‡] to the Ostrogoth king, Theodoric the Great.

Theodoric himself is said to have commissioned the work to validate the Goths' rule over the Roman world. Given that no history of the Gothic tribes survives in any form, one has to wonder about the nature of Cassiodorus's sources. What comes down to us, in the form of the *Getica*, is a fascinating but unreliable account which scholars still debate how far it can be trusted if at

---

*   More properly known as the *De Origine Actibusque Getarum*.

†   And we have to remember that, by the sixth century, the Roman Empire had been long divided, and was developing different traditions and styles of historiography.

‡   This translates as "master of offices" and was one of the most senior administrative posts in the Later Roman Empire. It is largely thanks to Cassiodorus's influence that the medieval monasteries of his time began to copy and retain the classical manuscripts which bequeath to us so much of our knowledge of the ancient world and form the cornerstone of Western art and culture.

all.* If it wasn't for Procopius, we would probably have to discount the *Getica*'s evidence for skiing in its entirety. But, taken together, these two references allow us to state that the act of skiing was known and practiced in Europe for well over a thousand years.† However, for many centuries, this knowledge remains confined to the far north, to Scandinavia and the northern parts of Central Asia, which includes Siberia.

At around the same time, in Tang Period China,‡ we also find references to "wooden-horse Turks who go hunting with wooden horses on their feet."§ Some two

---

* The issue is primarily that it is almost impossible to determine how much of the book is history and how much of it is myth, the two ideas being largely intertwined in the ancient world. Our ancient sources can often be littered with this conundrum. But one shouldn't be surprised: the Ancient Greeks, for example, considered the works of Homer to be history. Herodotus, the so-called Father of History, leaves us a book packed with myth and tall tales, though he is frequently good enough to point them out. Perhaps it is the human desire for the fabulous and fantastic that is in part of blame. One only has to look at ancient books to see why. On the one hand you have Arrian, the sober-minded historian whose *Anabasis of Alexander* remains one of our best sources for Alexander the Great's campaigns. On the other, you have *The Alexander Romance*, described by my old professor Thomas Wiedemann as "one of the great best sellers of the ancient world," which is packed with impossible and magical stories of its hero's achievements. And in the middle you have Quintus Curtius Rufus's *Histories of Alexander the Great*, which kind of interweaves the two.

† Not forgetting the prehistoric evidence, which leads us so much deeper into skiing's past.

‡ 618–907 C.E.

§ LeRoy J. Dresbeck, "The Ski: Its History and Historiography," *Technology and Culture*, vol. 8, no. 4 (Ocober 1967), page 474.

hundred years later, the Benedictine monk and historian Paul the Deacon described how the Lapps went "hunting on pieces of wood curved like bows."* The imagery of skiing begins to mount up in this period, the eighth century, too, with the Norse gods Ullr and Skadi frequently depicted on skis. And in the *Gastae Danorum*, written by the twelfth century Danish historian Saxo Grammaticus, we find our first reference to military skiing, when he tells us that the Finns used skis in combat both in advance and retreat.† We have to wait until 1556 to find the second one.

This evidence is frustratingly scant; as a general rule, our sources tend only to mention things done on skis rather than anything about the nature of the skis themselves. But, as with so many rules, there is an exception, which we can find in a twelfth century geography written by the Persian Sharaf Al-Zaman Tahir Marvazi, in which he describes the skis used by the Kim people of Central Asia. "If any of them wishes to go out to hunt the sable or the ermine or such like, he takes two pieces of wood, each three cubits long and a span wide, with one of the ends turned up like the prow of a ship, and binds them with his boots to his feet. In these he treads, rolling across the snow like a ship cleaving the waves."‡ But, though it frequently raises

---

* *Historia Langobardorum*, Book 5.

† *Gestae Danorum*, 9.

‡ LeRoy J. Dresbeck, "The Ski: Its History and Historiography," *Technology and Culture*, vol. 8, no. 4 (October 1967), page 475.

more questions than it answers, it allows us to be sure that
the act of skiing was practiced widely in the far north of
Europe and Asia for thousands of years. And that it was
of enough curiosity to the historians of the day to be worth
a mention, but not a digression.*

Skiing could perhaps have remained in this cul-de-sac
of practicality in the far north were it not firstly for the
Norwegians, and then for a Frenchman named Henry
Duhamel. In the years that followed Olaf Rye's epic ski
jump in 1808, it was the Norwegians who began the
practice of skiing as a leisure activity. When Captain Rye
made his jump, Norway was not an independent country.
Instead, it was a part of a union with Denmark that
began in 1397. The so-called Kalmar Union had originally included Sweden as well, but the Swedes broke
from it in 1521 and, in response, the Norwegians tried
to follow suit. Their rebellion failed, and Norway would
endure what nineteenth century romanticists referred to
as the "400 year night" of Danish rule.

This only came to an end as the Napoleonic Wars drew
to their close, when Denmark found it had been placed in
an invidious position. In 1807, with its army camped on
the Danes' southern borders, France attempted to force
an alliance between the two empires in the hope of using
the Danish navy to invade Britain, since the British

---

* You can bet, I'm sure, that if Herodotus had found a reason to write
  about skiing, there'd have been one helluva digression about it!

had effectively destroyed theirs at the Battle of Trafalgar in 1805.* In response, the British gave the Danes an ultimatum: ally with us, or surrender your navy and declare neutrality. Denmark did neither. So Britain attacked. On September 2, in what can only be described as a modern military experiment, the British fired rockets into Copenhagen, setting fire to the city. Just three nights later, the Danes gave up their fleet to prevent further bombardment.

In an effort to protect itself, the Danes allied with the French on October 31 of the same year. It did them little good: Britain blockaded their ports, plunging Denmark and Norway into an economic crisis that forced the former country into bankruptcy in 1813 and the latter into an unequal union with Sweden, which would last until 1905.

Throughout this period, the Norwegians tried to find ways to assert their independence. And, as the historian Andrew Denning puts it, they "pioneered leisure skiing in the mid-nineteenth century, and the sport rapidly became a marker of national identity that distinguished them from their Swedish overlords."† That is where the sport might have stayed, were it not for a French rock climber and mountain guide called Henry Duhamel.

Duhamel grew up in the small town of Gières in southeast France which, today, has been absorbed into the city

---

* In which John Franklin and Henry Pelgar, from chapter 1, both served.

† Andrew Denning, "How Skiing Went from the Alps to the Masses," *The Atlantic*, February 23, 2015.

of Grenoble. He was, by all accounts, a sickly child, so his doctors recommended he spend more time in the Alps in the hope that the region's bright fresh air might help him get better. It did more than that: not only did his health improve, his time there left him with a lifelong love of the mountains and of climbing.

For a young alpinist from Grenoble—no doubt possessed of the same drive to climb as the 1920s British mountaineer George Mallory who, when asked why he wanted to climb Everest, famously replied, "because it's there"—there was one nearby and unsummitted mountaintop that had to be conquered: the 3,983 meter peak of the Grand Pic de la Meije in the Massif du Haut-Dauphiné.*

Over the three years between 1875 and 1877, Duhamel tried the climb from all sides of the mountain, to no success. On the second attempt, on the south face, which from afar looked like it might offer a route to the top, he and his companions François Simond and Edouard Cupelin came upon an almost sheer wall of rock at 3,500 meters.† They built a cairn to mark how far they'd climbed and went no further, with Duhamel claiming in a letter to Adolphe Joanne, president of the Club Alpin Français (CAF): "I will permit myself to state with assurance, la Meije cannot be climbed."‡

---

* Located in the Dauphiné region of Southeastern France, home of the fantastic French potato dish, Pommes Dauphinoise.

† Claude Gardien, "Inaccessible," *The Alpinist*, September 14, 2017.

‡ Ibid.

He was wrong. For, on August 16, 1877, fellow CAF member Emmanuel Boileau de Castelnau and his guide Pierre Gaspard reached the top. "Ah," said a disappointed Duhamel, "So it's done." He would never attempt the Grand Pic de la Meije again.

He didn't give up on the mountains, continuing to climb throughout 1878, when he discovered a completely new way, to him at least, to approach them, in Paris, of all places.

The *Exposition Universelle** was held at the Palais du Trocadéro from May until November that year. Duhamel went along in search of snowshoes, which he hoped to find at the Canadian stands. Those he had found in Europe had proved to be unsatisfactory, leaving him unable to explore the mountains as he'd like each winter. While he found his Canadian snowshoes, he also bought something else at the Swedish–Norwegian stands: skis.

This was the first pair of skis ever to be deployed in the Alps, and they came with one small drawback: Duhamel had no idea how to use them.[†] He set himself a course of research and experimentation, learning first how to crash and the obvious lesson that it is much quicker to come down a mountain on skis than it is to climb up.[‡] It would

---

* Paris's third World Fair.

† They were without instructions . . . http://pistehors.com/back country/wiki/Isere-Drome/Chamrousse-Ski-History.

‡ Ibid.

not be for another eighteen years, when two Norwegian army officers visited Chamrousse, where Duhamel practiced and encouraged others to join in, that any of the local enthusiasts would learn how to turn on them.

By this time, Duhamel had perfected the use of bindings to keep his skis firmly on his feet and had traveled, in 1890, to Finland to learn skiing secrets from the local Sami people. He returned that year for fourteen pairs of new, Norwegian-made skis, *pour encourager les autres*. And, with the formation of the French Alpine Ski Club, constituted on February 1, 1896, Alpine skiing was born.

Duhamel's introduction of the ski transformed Alpine life beyond recognition. Before its arrival, even the most basic social or economic interactions between Alpine villages in winter was almost impossible. As LeRoy J. Dresbeck wrote in his 1967 article, "The Ski: Its History and Historiography," "The depth of the snow in the mountains provided three basic dangers to travel: cold, fatigue, and avalanche."* So people stayed home. Dresbeck goes on: "In mountain terrain sleighs and sledges were impractical, and the snowshoe was too slow and tiring to be of much use in heavy snow."†

The ski changed this. Between 1899 and 1905, its use began to spread, changing people's lives dramatically.

---

* LeRoy J. Dresbeck, "The Ski: Its History and Historiography," *Technology and Culture*, vol. 8, no. 4 (October 1967), page 467.

† Ibid.

The ski offered freedom from the jail of winter. Suddenly things the rest of the lower lying regions of Alpine countries enjoyed became possible: postal services, access to doctors, or, as in the case of the entire village of Val d'Isère in 1913, attending Midnight Mass on Christmas Eve.* Thanks to the gradual and then snowballing† popularity of winter sports, an entire new economy based on skiing, skating, and bobsledding became possible. There was just one thing holding it back: how do you get skiers up a mountain?

Now,‡ I'm no skier. I have skied twice on mats, at a UK version of summer camp, and been on a single weeklong skiing holiday where, despite classes, I realized that I didn't so much ski as plummet. But I learned that digging your skis' edges into the snow to side-step up the nursery slope before snowplowing back down is a woefully inefficient way to climb a hill. I mean, there's just no way that the athletes at the first Winter Olympics, held in Chamonix in 1924, some eleven years before the invention of the ski lift, were side-stepping their way up the mountain side to compete in each event. There has to be a better way.§ And there is.

---

* LeRoy J. Dresbeck, "The Ski: Its History and Historiography," *Technology and Culture*, vol. 8, no. 4 (October 1967), page 467.

† Pun intended . . .

‡ Again, please forgive the anecdotage.

§ Imagine all those athletes, side-stepping their way to the top of the downhill run—it's preposterous.

Today, it is known as "uphill skiing," "skinning," or "Alpine touring." It has evolved into something of an extreme sport, allowing its practitioners to explore a resort's mountain away from the prepared pistes.

The name "skinning" is a holdover from how this was originally done, by wrapping your skis with animal skins to provide grip as you work your way uphill.* You also need a telemark or AT binding which frees your heel from the ski to offer a greater range of movement. And, while it's much more efficient than ski-class side-stepping, it is hard cardiovascular work, which means it's not everybody's cup of tea. The surge in skiing's popularity, which saw the sport spread from the Alps to North America, Chile, and Japan—which enjoyed a serious skiing craze in the 1930s—was driven more by the exhilaration of speed provided by Alpine skiing than by the sheer hard work of the Nordic cross-country style.

It's hardly surprising that people began to look for a more mechanical solution to ferry skiers up the mountain. What is rather more surprising is who found it: an employee of the Union Pacific Railroad.

In 1935, the Railroad's chairman, William Averell Harriman, cottoned on to the growing popularity of winter sports. He decided to create the ultimate mountain resort which would, of course, be serviced exclusively by his railway. His first order of business was to decide

---

* Today, skins have been replaced by nylon felt.

where to put it. To this end, Harriman commissioned an Austrian count, Felix Schaffgotsch, to find the perfect location.

Count Felix searched the West, traveling through Washington State, Utah, and Northern California before he finally found a sheltered site in the Sawtooth Mountains, some hundred miles from Boise, Idaho. The Sun Valley Resort would become the first destination ski resort in the United States. Per Harriman's specifications, it was to have everything its visitors could possibly desire: high-end accommodations, fine dining, swimming pools heated by hot springs, and, of course, skiing. For this last attraction, Harriman wanted to find a way to broaden the pastime's appeal beyond the hardcore uphill skiing athletes to anyone who wanted to try it. Which meant he and his team had to find a less exerting way to bring their guests up the mountain. It was quite the conundrum, and one that Harriman was determined to solve, writing in a telegram on April 15, 1936: "It is essential to develop a method of lifting skiers 2,000 feet above the valley floor."*

One of the Railroad's bridge designers, a young, largely self-taught civil engineer called James Curran, had the answer. Prior to joining the Union Pacific, Curran had held a variety of jobs, including a stint as a pool hustler.

---

* Daniel Engber, "Who Made the Ski Lift," *New York Times Magazine*, February 23, 2014.

But it was his time working for the United Fruit Company that held the key. There he had designed a contraption to load bananas onto ships in Honduras. A wire cable with hooks suspended from it at discrete intervals ran from dock to deck, allowing stevedores to efficiently shift bales of bananas aboard, hooking them and unhooking them from the cable as required. Curran proposed that, if one replaced the fruit hooks with chairs, the same kind of system could move people as well.

Curran's colleagues were not convinced. In fact, the general consensus was that what was good for bananas would be actively dangerous for people. But Curran had a plan to demonstrate the safety of his design, building a wooden frame over the back of a pick-up truck from which he could suspend a chair about half a meter from the truck's rear wheel.

In the summer of 1936, he conducted a testing program in a rail car repair shop in downtown Omaha, Nebraska. And brave volunteers, wearing roller skates—since the weather ensured they couldn't try it out on skis—rode the chair until they had determined the ideal speed for the system's future riders to get on and off in safety. The chair lift was born.

Before Curran's invention appeared at ski resorts around the world, skiers hadn't only "skinned" to climb the mountains. Some new resorts had installed rack railways to shuttle people to the top. Other areas had used horses and sleighs. But neither could shift the hundred people every hour that Harriman required.* Tow ropes and primitive versions of the T-Bar, which is still in use today, had also been tried. None of them offered the kinds of views over the valley, nor the sense of anticipation from seeing the slope on which you'd soon be skiing from above, that Curran's machines could provide. When Union Pacific opened Sun Valley in December 1936, his chairlifts were among its star attractions, carrying guests uphill at a stately four miles per hour.

It wasn't just skiing's popularity that encouraged Harriman to invest in Sun Valley, chairlifts, and the rail infrastructure needed to supply the resort with its guests and its various needs. He was also aware that his railroad, and the others that spanned the United States, did not simply transport people from city to city for work. It also offered them adventure and relaxation. Businesses like his were at the forefront of a new and burgeoning leisure industry that was both transforming society and, in the case of Sun Valley, the environment.

Even in 1936, this was not a new idea. In fact, it was one that the railroads had been pushing for over thirty years. In 1905, Fred Harvey† had, with the help of the Atchison, Topeka and Santa Fe Railway, for whom he had built his business, opened the celebrated El Tovar hotel on

---

*   Frank V. Persall, "Who Invented the Ski Lift? An Engineering Masterpiece," Snowgaper.com.

†   Of Harvey House fame.

the south rim of the Grand Canyon. Built by the railway's chief architect, Charles Frederick Whittlesey, at the end of a specially commissioned rail spur, the hotel was, like Sun Valley, a destination resort that offered luxury in the midst of the wilderness.[*]

Men like William Harriman and Fred Harvey were reimagining and reshaping remote environments for people's pleasure in a way that had never happened before. Certainly not on such a scale. While hotels had been built to meet the needs of eighteenth century Grand Tourists, or even those of the users of Pausanias's second century guidebook to Ancient Greece, the oldest surviving tome of its kind, no one had created such places in the middle of nowhere, purely for leisure, in any other era.

Like the ice in our fridge-freezers or the out-of-season strawberries in the supermarket, we don't really think about the extraordinary luxury of things like the weekend. Nor do we realize how recent the concept of it is, nor even that the idea of the seven-day week has never been set in stone.

It is generally estimated that the seven-day week was a Judaic idea originating to the sixth century B.C.E.[†] But throughout history, this has been far from standard. The

---

[*]   You can still visit the El Tovar today.

[†]   Eviatar Zerubavel, *The Seven Day Circle: The History and Meaning of the Week*, University of Chicago Press, 1989.

Ancient Romans had an eight-day week and, on the eighth, or the *nundinae*, children didn't have to go to school and farmers took a day off from their fields to sell their produce at market. Meanwhile, the French Revolutionary Calendar imposed a ten-day week upon its citizens.

As such, the week is an entirely arbitrary, man-made invention. As the writer Witold Rybczynski has pointed out,* a full day is determined by the (roughly) twenty-four hours between sunrises, a month was originally measured by one full orbit of the moon, a year by the earth's orbit of the sun or, before Galileo, one full cycle of the seasons. But the seven-day week is determined by nothing but us, though it has been suggested that it evolved from breaking a twenty-eight-day lunar cycle into four seven-day units.

The weekend is just the same. But its association with time off is not indelibly tied, as you'd expect, to the concept of the sabbath, which falls in the three Abrahamic faiths—Islam, Judaism, and Christianity—on Friday, Saturday, and Sunday respectively. The word for it doesn't appear in English in common usage until 1878, though it had existed in the north of the country since the 1630s, referring to the period between Saturday afternoon and Monday morning. This connection to the sabbath does not tie said day off to leisure. Far

---

*   In his excellent article, "Waiting for The Weekend," published in *The Atlantic*, August 1991.

from it: it creates a day for religious observance and the absence of work, any work, depending upon how deeply one practices it—just think of Eric Liddell, the British runner who famously refused to compete because his race was to take place on a Sunday, as depicted* in the film *Chariots of Fire*. Its invention, in the late nineteenth century, was British.†

Prior to this, the British work week ended on a Saturday evening. But, as the country became decreasingly Puritan,‡ this weekly holiday became, in Rybczynski's words, "a chance to drink, gamble, and generally have a good time."§

By the time of the Georgian era in Britain, as the country became increasingly prosperous, workers were, for the first time in hundreds of years, being paid more for their labor than they needed to just survive. This meant they could buy things, and they could pay to have fun. Or they could choose to do more work and earn more money. Most did not.

This was an era that didn't really need such a thing as a formal weekend. Not only were the usual religious holidays observed, complete with time off, but should

---

* Albeit not with complete accuracy . . .

† Rybczynski Ibid.

‡ Such a dour bunch, who prohibited the very idea of fun on the Lord's Day.

§ Ibid.

some touring attraction come to town—a fair or a circus or traveling zoo—people downed tools and went to see it. Writes Rybczynski:

> The idea of spontaneously closing up shop or leaving the workbench for the pursuit of pleasure may strike the modern reader as irresponsible, but for the eighteenth-century worker the line between work and play was blurred. Many recreational activities were directly linked to the workplace, since trade guilds often organized their own outings and had their own singing and drinking clubs and their own preferred taverns.

The Industrial Revolution changed that. After all, you cannot operate a factory if everyone on your production line buggers off because someone's brought a zebra to town. Mass production requires order and a constant working week, and that in turn requires regular time off for workers to recover from their toil.

Sunday remained a holiday from work. But, unlike today, it was not Saturday that began the not-yet-dubbed weekend, but Monday that ended it. This may have started so that people could recover from Sunday's drunkenness, or because, all too often, public events were banned from happening on the sabbath, but several trades, including miners and weavers, took the day off on the Monday after payday (usually a Friday or a Saturday). In fact, the practice became so common that the day became known as Saint Monday. And, by the mid-nineteenth century, there were plenty of leisure entrepreneurs ready to cash in on it.

Monday became the day for horse races or cricket matches. On Monday, attendance at botanical gardens soared. On Monday, working class people went out dancing or to the theater.

This creates what Rybczynski describes as a curious contradiction. It wasn't so much that these entrepreneurs were commercializing people's newly discovered leisure, but that people were discovering leisure thanks to the entrepreneurs' commerce. He goes on, arguing that, "Beginning in the eighteenth century, with magazines, coffee houses, and music rooms, and continuing throughout the nineteenth, with professional sports and holiday travel, the modern idea of personal leisure emerged at the same time as the business of leisure. The first could not have happened without the second."

But Saint Monday was not to last. Thanks to protests from religious groups, who argued that both the Monday day off and the dissipation and drunkenness of Sunday dishonored God's day, and by other social reformers, attitudes shifted. By the end of the century, factory owners had begun to end work halfway through Saturday and beginning it again on Monday morning. It was in their interests to do so: a six-day working week would lead to so many workers not showing up that they'd have to close for the day anyway. The Saturday half-day seemed like a good compromise.

And by the 1870s, phrases like "spending the week end" began to enter the lexicon.*

Saturday afternoons quickly became a time for visits to the park, watching or playing sport—it's no accident that, initially irregular, football fixtures began to take place on Saturday afternoons, leading to the creation in England of the world's first football league in 1888†—or an afternoon at the local pub. Saturday nights became the ideal time to visit the theater or the music hall. It codified, especially among the working classes,‡ what Rybczynski describes as "something new: the strict demarcation of a temporal and a physical boundary between leisure and work."§

---

* Though not among the upper classes, as a friend reminded me recently. To say "Are you coming for the weekend?" would be common. Instead, receiving an invitation to somewhere impossibly grand, like Chatsworth in Derbyshire, you'd be asked, "Are you coming from Friday to Monday?" By such linguistic tricks do we Brits derive our snobbery and tell each other apart.

† The idea for it was hatched by William McGregor, a director of Aston Villa Football Club. It was initially comprised of just twelve teams rather than the ninety-two we have today, spread across several divisions. At the initial meeting, held to create it on March 23, 1888, McGregor suggested calling it "The Associated Football Union," but this was felt to be too close to the "Rugby Football Union." So William Sudell, representing Preston North End, suggested "The Football League," and the name quickly prevailed.

‡ Aristocratic leisure activities remained very much of the "hunting, shooting, fishing" variety they'd always been.

§ Ibid.

It was an idea that would spread around the world. It did so not so much at the instigation of a burgeoning union movement—though unions certainly played their part—but at the instigation of the bosses. And, in the United States at least, chief among those bosses was the car manufacturer Henry Ford. It may seem counterintuitive, but this was all a ploy to make the Ford Motor Company more profitable. As Grant Oster writes on the website *Hankering for History*,* by 1913 the business was hemorrhaging money.

Ford had already had two businesses go down on him. His first, the Detroit Automobile Company, which he'd founded in 1899, folded by 1901. His second, the Henry Ford Company, failed when he fell out with his investors. And now the third was in trouble too, just ten years after its creation, in spite of the success of the Model T, which was first produced in 1908, and the creation of the world's first assembly line to build it.

Ironically, it was the very success of the Model T that had led to Ford's problem. The concept of the assembly line, introduced into the factory in 1913 and leading to the production of 202,667 new cars that year, was tough on his workers, leading to a high turnover in their numbers, even as the assembly line meant that he required fewer of them.

---

* https://hankeringforhistory.com/history-weekend/.

Ford's solution, which he instigated on January 5, 1914, was to more than double their wages, from $2.34 to $5 a day. Not only that, he changed their hours, implementing the eight-hour, five-day working week we know today.

These compassionate innovations proved to be staggeringly successful. Worker retention was no longer a problem, which in turn dropped the costs of training new staff. And, of course, with more money in their pockets, Ford workers could now afford one of the Model T's they'd helped to build.

It would not be long before other manufacturers, noting Ford's growing success, would follow suit.

Even so, it would take until the 1930s for the idea to catch on more widely. Which makes William Harriman's opening of the Sun Valley Resort in 1936 perfectly timed for his clientele, now with more leisure time on their hands and money in their pockets, to take a holiday.

Holidays, it turns out, are a much older phenomenon than weekends, with the Ancient Romans being the first people to be able to travel for fun, and for one key reason: the Empire had certain facilities to make it possible. Firstly, it built good roads. Secondly, its famous army guarded secure borders and did away with banditry—well, mostly, there were certain parts of mountainous southern Turkey you'd have been wise to avoid. And thirdly, its

not-so-famous navy had wiped out piracy in the Mediterranean by the end of the first century B.C.E.[*]

The Romans had quite the infrastructure for holi-
daying: inns, restaurants, local tour guides, even guide-
books, though only one survives. Pausanias's book,[†]
written in the second century C.E., was a guide to Ancient
Greece and came in ten scrolls, detailing everything
you could possibly want to know about the place, from
geography to art, architecture, and culture. But the years
leading up to the fall of Rome, and indeed the fall itself,
put a stop to recreational travel. Once again, it became
too dangerous. But, as Europe put the dark ages behind
it, travel re-emerged in the form of pilgrimage, with many
of those storied routes, like the Camino de Santiago, still
being walked today.[‡]

---

[*] Though not before a young Julius Caesar was kidnapped by a band
    of Cilician pirates while he was on his way to Rhodes in 75 B.C.E. As
    Plutarch tells it, he negotiated his ransom up to reflect his impor-
    tance, generally bossed them around like he was their commander,
    and then, when the ransom was paid and Caesar released, raised a
    navy in Miletus, sailed off, and captured the lot of them, all without
    any previous military experience. Finally, when the Governor of
    Asia dithered over their punishment, Caesar personally had them
    crucified.

[†] Mentioned briefly, above.

[‡] I mention this largely to plug Emilio Estevez's moving and
    delightful film, *The Way*, simply because it features one of my
    favorite Martin Sheen performances.

On these pilgrimages, the ambitious, dedicated traveler could cover much of the continent, as evidenced by the Benedictine monk Matthew Paris's thirteenth century *Map of the Itinerary from London to Jerusalem*. Alternatively, if a pilgrimage was not for you, you could, like Marco Polo, travel in the company of merchants, if only for safety in numbers. But this kind of travel takes up a lot of time, time unavailable to the kinds of customers William Harriman wanted to attract.

By the Tudor era, travel was largely the preserve of royalty. As the Renaissance dawned, it was increasingly practiced by traders and explorers. And it was only by the eighteenth century that people began to travel for pleasure once more. But, for those Grand Tourers, it was probably much more difficult than it had been for the Romans, even as a new infrastructure grew up to meet them. So it was up to the railways, like the one owned by Harriman, to reintroduce the concept to the masses.

The railways took away the time it took to get from one place to the next. When Britain introduced the concept of the Bank Holiday—a three-day weekend where you get Monday off as well—the railways capitalized on selling tickets for three-day holidays, often to the seaside, booming the economies of towns like Blackpool, Skegness, Brighton, and Bognor Regis, to name a few, where visitors could enjoy many of the things we still today associate with the British seaside.

Ice cream.* Fish and chips.† Seaside piers and Punch and Judy shows.‡

While trains could take the working and middle classes for a seaside break, steamships opened even more possibilities for the wealthy. Suddenly, if you could afford it, you could visit Paris, Cairo, or Shanghai, obtaining local currency with a "circular note" from your bank—the forebear of the traveler's cheque.§

You could even take the Orient Express to Istanbul.

By the early twentieth century, the combination of affordable cars, transcontinental train networks, better labor laws, and more money in people's pockets would ignite the leisure industries in both North America and Europe, allowing travel and holidays to become a very big deal. Even the Great Depression wasn't enough to turn back the tide. Indeed, by 1933, the legislation passed to

---

*   See chapter 3.

†   Which turns out to be far from a British invention. The practice of frying fish was imported to Britain by Jewish immigration. Thomas Jefferson famously wrote of frying fish "fried in the Jewish fashion" on a trip to England. The arrival of the chip is harder to pin down. But the first fish and chip shops appeared in 1860. Most food historians agree that the first was opened by a young Ashkenazi Jewish immigrant named Joseph Malin, and his establishment was so successful that it remained in business into the 1970s.

‡   Not so much anymore, thanks to the violence, but there were still plenty around when I was small.

§   Post the Romans and prior to this, the only people to offer international banking, particularly to pilgrims, were the Knights Templar.

bring about the First New Deal had broadly stabilized
the American economy, and the laws associated with the
Second were beginning to make their way through Congress. So William Harriman's bet on the Sun Valley Resort
and winter sports just two years later* was very much on
the money.†

Meanwhile, in Europe, it is also in the 1930s that
we see the first winter chalet holidays being offered to
the public. In 1932, a Jewish-Austrian businesswoman
called Erna Low placed an advertisement in the *Morning
Post*, offering a package skiing holiday for £15.‡ The deal
included travel to and from the chalet, food, accommoda-
tion, ski hire, lessons, and even a lesson in German.

Low was a former Austrian javelin champion who had
moved to Britain the previous year to study for a PhD,
financing herself by working as a language tutor. The
tutoring was not bringing in enough money to make trips
back to Austria to visit her mother, nor to go skiing. The
chalet rentals could provide it.

Low is the first person we know of to think of this brilliant idea. Back in the day, chalets in the European Alps
were only ever used in the summer. Farmers would herd

---

* The idea was conceived in 1935 and put into practice in '36.

† He wasn't alone in making a bet on holiday resorts that year. In
1936, Billy Butlin created the first of his affordable Butlin's holiday
camps in 1936, which was opened by the famous aviatrix Amy
Johnson.

‡ Approximately $1,140 today.

their cows up to them when the snows cleared, then they'd graze and milk them to make cheese before returning to homes further down the mountain for the winter.

With all these properties standing empty in the middle of prime ski slopes, Low realized that, by renting and subletting them, she could make a tidy profit even with all her other expenses on top. Although the chalets of this era were fairly basic affairs—there was often a lack of hot water, you had to share the bathroom, and sometimes even the bedrooms—her idea was a roaring success. So much so that her business, Erna Low Ltd, survives to this day, specializing in skiing trips and Alpine property. Low herself is considered by the travel industry to be one of the pioneers of the package holiday.

It was pioneers like Duhamel and innovative entrepreneurs like Harriman and Low who lay behind the boom in winter sports in the pre-War era. But it wouldn't be until after the War, when the costs of international travel began to fall, allowing more and more people to visit the countries of their dreams, that the sport would really take off.

On October 21, 1968, filming began on the much anticipated sixth installment of the James Bond franchise, *On Her Majesty's Secret Service*, in the Swiss canton of Berne, not far from the village of Mürren.

Mürren already held a significant place in the history of skiing. For it was here, in 1922, that one of the sport's leading pioneers, Sir Arnold Lunn, invented the discipline of slalom racing. In his opinion, at the time, the sport needed something to test a skier's ability to turn while losing as little speed as possible, one where, in his words, "a fast ugly turn is better than a slow pretty turn."

Slalom had previously existed, but before Lunn it was a matter of style over speed. Lunn introduced the race against the clock. In laying out the course for this first race, held on January 21, 1922, he placed the flags to ensure specifically that all the main styles of turning would be required.[*]

The village would find itself at the epicenter of skiing's explosion into the international consciousness.

The early Bond films, in particular, always contained within them a not-very-subtle subtext of aspiration—the cars, the watches, the women, the exotic locations: all were designed to provoke the audience's desire for the things they couldn't have, the places they couldn't visit. The Jamaica of *Dr. No*[†] or the Bahamas of *Thunderball*, to name but two.

Now it was the turn of the glamorous Swiss Alps to shine in the light of Bond, as captured by its cinematographer Michael Reed. Handheld cameras operated by the skiers themselves put the audience deep into the action, while cameraman Johnny Jordan developed a rig which allowed him to be dangled on an adapted parachute

---

[*]   The race was won by an Englishman called J. A. Joannides.

[†]   My mother well remembered the excitement on the island when the Bond team came to town. Back then she worked for British West Indian Airways in Montego Bay, and was friends with Marguerite LeWars, who played the small role of Annabel Chung in the movie.

harness some five-and-a-half meters below a helicopter, allowing him to shoot on the move from any angle, all the better to capture the action's excitement and exhilaration.*

Released in 1969, *On Her Majesty's Secret Service* served as the second of a one-two punch that landed skiing firmly in the public mind. The first was thrown in Grenoble which, between February 6–18, hosted the tenth Winter Olympics. It was the first to be widely televised, thanks to the rapid advance of television satellite technology.

The 1956 Cortina d'Ampezzo Winter Games were the first ever to be broadcast, albeit in a limited fashion via the Eurovision network, which meant that only nine European countries had access to the feed. But at least it was live. Four years later, CBS bought the rights to show the 1960 Games, held in Squaw Valley, California, which made sense since they were taking place in the US.† Meanwhile, the 1964 Games, held in Innsbruck, were the subject of a feature-length film documentary.‡ But the TV footage was all on black-and-white tape, and none of it was live. ABC, who had by now acquired the

---

* I saw *OHMSS* on the big screen not too long ago, and these sequences remain absolutely mind-blowing.

† During one of the men's slalom races, officials noticed that one of the skiers might have missed a gate on his descent. They asked CBS if they could review the tapes. It was a moment that inspired the network to create "instant replay" in the years that followed.

‡ Which, if you're so inclined, you can watch here: https://olympics .com/en/original-series/episode/innsbruck-1964-official -film-ix-olympic-winter-games-innsbruck-1964.

US broadcast rights, aired some sixteen-and-a-half hours of footage, but very little of it went out in primetime.

The 1968 Winter Games in Grenoble would be different. Firstly, its coverage was in color.* And, while only a couple of events went out live—the opening ceremony and the ladies figure skating final—each event was shown on the day it happened.

ABC put out twenty-two hours of coverage over eleven days to significant critical acclaim.† With a massive 94.6% of American households now owning TVs, broadcasting just three channels, the Grenoble Games were a ratings hit. So much so that they turned the French skier Jean-Claude Killy into a major US star.

Killy had been skiing competitively since he was sixteen, but his early career was blighted by a distinct lack of success. He was fast—damned fast—but he had a tendency to crash. A lot. Not only that, his 1964 Olympics, in which he was entered in three events, were destroyed when diseases he'd picked up during his national service in the French Army in Algeria came back to haunt him. But in 1967, he hit peak form. At Grenoble, he completed the unprecedented feat of taking the gold in the slalom, giant slalom, and downhill events.

---

*   With the exception of a couple of events taken from the black-and-white French feed.

†   "The World Comes Together in Your Living Room: The Olympics on TV," https://jeff560.tripod.com/tv8.html.

He became a sensation. He had the charm, the looks, the Gallic insouciance. Endorsements quickly followed, not only with the ski manufacturer Head Ski, but with massive US brands. He made advertisements for American Express. He became a spokesman for United Airlines, Schwinn bicycles, and, perhaps most notably, for Chevrolet.

Suddenly, here was a French skier, alongside the recent Heisman Trophy winner and (then) hugely popular O. J. Simpson, selling American cars to Americans. Skiing had broken into the mainstream, and Jean-Claude Killy was its face. So much so that he was the subject of a profile by Hunter S. Thompson for *Scanlan's Monthly*, in which the writer followed Killy on a Chevrolet promotional tour.

Thompson, in the article, would note that, while Killy worked just as hard on pitching as he had on his skiing, he possessed a certain Gatsby-like quality, writing of him,

> That description of Gatsby by Nick Carraway— of Scott, by Fitzgerald—might just as well be of J.-C. Killy, who also fits the rest of it: "Precisely at that point [Gatsby's smile] vanished—and I was looking at an elegant young roughneck, whose elaborate formality of speech just missed being absurd . . ."
>
> The point is not to knock Killy's English, which is far better than my French, but to emphasize his careful, finely coached choice of words. "He's an amazing boy," I was told

later by Len Roller. "He works at this [selling Chevrolets] just as hard as he used to work at winning races. He attacks it with the same concentration you remember from watching him ski."*

In the article, Thompson describes one automotive journalist pondering just why Chevrolet had signed Killy up. After all, O. J. Simpson was a much bigger draw. But Thompson got it:

> Then I looked at the crowd surrounding Killy. They were white and apparently solvent, their average age around 30—the kind of people who could obviously afford to buy skis and make payments on new cars. O. J. Simpson drew bigger crowds, but most of his admirers were around twelve years old.
>
> Mark McCormack signed to manage Arnold Palmer a decade ago—just prior to the Great Golf Boom. His reasons for betting on Killy are just as obvious. Skiing is no longer an esoteric sport for the idle rich, but a fantastically popular new winter status-game for anyone who can afford $500 for equipment.

---

* Hunter S. Thompson, *The Great Shark Hunt: Gonzo Papers Vol. 1*, Simon & Schuster, 1979.

Five years ago the figure would be three times
that, plus another loose $1,000 for a week
at . . . Sun Valley, but now, with the advent
of snowmaking machines, even Chattanooga
is a "ski-town." The Midwest is dotted with
icy "week-night" slalom hills, lit up like the
miniature golf courses of the Eisenhower age.*

Television brought skiing to the world. In Britain,
the program *Ski Sunday*, which in 1978 was spun out
of the BBC's coverage of the 1976 Innsbruck Winter
Games, drew a regular audience of millions[†] and made
household names of skiers like the Austrian Franz
Klammer and Canadian Steve Podborski,[‡] all while
prompting viewers to make annual pilgrimages to the Alps
to ski, drink glühwein, and dine on dishes like fondue[§]
or tartiflette.

Skiing evolved into a massive business, providing
exhilaration and relaxation to millions each year. But, as
ever, we have to ask: at what cost?

At first glance, skiing would seem harmless, a cel-
ebration of our glorious mountain environments. But the

---

*   Ibid.

†   Again, in a three-channel environment.

‡   Though not quite to the level where they could match Jean-Claude
    Killy for endorsements . . .

§   The subject of a massive food craze in 1970s Britain.

industry faces a number of problems, not least because, today, winters are at least a month shorter than they were thirty years ago.[*] While many resorts across the world are working hard to make themselves eco-friendly, their footprint on the environment is significant.

Among the issues they face, one of the most significant may be the need to make artificial snow. According to some estimates, it takes about 4,000 cubic meters of water to cover just one hectare of piste for a skiing season.[†] Not only that, artificial snow takes longer to melt than its natural counterpart, which prevents early germinating mountain plants from pushing through the thawing earth. Furthermore, artificial snow is four times harder than natural snow,[‡] and it's expensive to produce. Since the water used is often drawn from nearby rivers, it pumps nutrients into an ecosystem where they don't belong, disrupting high altitude biodiversity.[§] Its water use often places artificial snow's needs in direct competition with the provision of drinking water, so much does it consume. One third of resorts in the French Alps now face drinking water

---

[*]   Ellie Ross, "Hot Topic: Can Skiing Ever Be Green?," *National Geographic*, April 8, 2019.

[†]   Leo Hickman, "Can you ski and be green?," *Guardian*, September 28, 2008.

[‡]   Ellie Ross, "Hot Topic: Can Skiing Ever Be Green?," *National Geographic*, April 8, 2019.

[§]   Leo Hickman, "Can you ski and be green?," *Guardian*, September 28, 2008.

shortages. At one Massachusetts resort, the Wachusett Mountain Ski Area, snowmaking can use almost 20,000 liters a minute.[*]

Deforestation, performed in ski resort creation, also affects the biodiversity of any chosen area, particularly in North America, where sites are often placed in forested regions.[†]

But the big issue is power consumption, even if we don't consider the carbon footprint of one's flight to a resort. A single ski lift, powered by electricity, uses the same amount of power as almost four households in one year.[‡] Snowmaking uses some 25,000 kilowatt-hours to cover just one hectare of piste each season.[§] That's before one even considers what it takes to heat and light each one.

Yet the attraction of winter sports in general, and skiing in particular, remains. It lies in the exhilaration of its speed, in wonderful days in the outdoors, surrounded by the spectacular beauty of the mountains. It lies in the challenge itself. Skiing is, like golf, one of those sports you don't have to be particularly good at to enjoy. Even as a spectator, watching an event, be it skiing, bobsledding,

---

[*] Frederic Beaudry, "Ski Resorts and Their Impact on the Environment," Treehugger.com, November 3, 2019.

[†] Ibid.

[‡] Ibid.

[§] Leo Hickman, "Can you ski and be green?," *Guardian*, September 28, 2008.

or figure skating, the thrill lies not just in the competition but the challenge. When we watched an athlete like Usain Bolt compete at the peak of his career, the result is never in doubt. But the joy of it lay in our opportunity to glimpse the impossible. To see him go for the record; to see him challenge time itself. So it is in all sports, even mountaineering. When Jean-Antoine Carrel and Jean-Baptiste Bich made the first ascent of the Matterhorn on July 14, 1865, it was front-page news around the world. Because of the challenge.

So the question, for those who practice these things at the highest level, becomes: "Can I ski the Cresta Run?" "Can I climb the north face of the Eiger?" "Can I, a mere human being, beat my environment?" No longer are these cold places somewhere we struggle to survive. They have become ours to conquer. The hidden consequence is that we haven't conquered them in the way that we perhaps intended.

8

The Vanishing

On August 22, 2018,* and to very little fanfare, a container ship called the *Venta Maersk* put out of Vladivostok with a cargo of Russian fish and South Korean electronics, bound for St. Petersburg. It is a voyage that ordinarily covers 14,600 nautical miles and takes about two months. She did it in thirty-eight days.

Instead of plotting the usual course, which takes one south to the East China Sea, across the Indian Ocean and through the Suez Canal, the *Venta Maersk* sailed through the Arctic.

Hers was the very first cargo to sail the so-called Northern Sea Route, and took place almost exactly 140 years after the route had first been navigated by the Finnish-Swedish explorer Adolf Erik Nordenskiöld in 1878–9.

---

* https://safety4sea.com/venta-maersk-calls-at-st-petersburg
-concluding-passage-of-northern-sea-route/.

Perhaps the most remarkable thing about it is that nobody seemed to notice. Of course the story made the news,* but it was hardly front-page, above-the-fold stuff. Given the sheer quantity of newspaper ink expended on polar exploration in its golden age, and after all the costs and casualties racked up over centuries to chart both the Northeast and Northwest passages, you'd have thought it might merit a little bit more of a splash.

You might also think that here was an almost perfect, simple, visual way to demonstrate to the wider public that climate change is very, very real. This is something that matters greatly to those advocating our need for change in the face of a warming world. For, while the climate crisis has been in the public consciousness solidly since the 1980s, there has been no galvanizing image or idea to drive home with immediacy the need for action.

As little as two years before the *Venta Maersk* voyage, everyone knew that the Arctic ice was melting. Images of hungry polar bears on dwindling ice floes had seen to that. The Maersk company itself didn't think that such a voyage would be possible for at least a decade. By 2020, some 32.97 million tons of cargo were shipped along the NSR, while in February 2021, the Russian gas tanker *Christophe de Margerie* completed the trip in winter. Another first.

That this has happened so fast comes as no surprise to Arctic scientists. They've been warning us for years that the

---

* I found it on Reuters.

sea ice was retreating on a bell curve or, to put it another way, in much the same way that, in Hemingway's *The Sun Also Rises*, Mike Campbell went bankrupt: gradually, and then suddenly. We are arriving at a moment when what once seemed purely theoretical, and could thus be easily ignored, now lies before us: the world's natural ice is vanishing. We know why. We have the means to do something about it. And yet we lack the will.

But, for a brief moment in the late 1980s and early '90s, it seemed as though we might.

In May 1985, a team of scientists from the British Antarctic Survey published an article in *Nature* detailing the depletion of ozone in the upper atmosphere over the continent. They hypothesized that this had been caused by the release of chlorofluorocarbons (CFCs), which were widely used in aerosol cans and refrigeration. In doing so, they were drawing from work by Frank Sherwood Rowland and Mario Molina, who had been studying the effects of CFCs upon the atmosphere since 1973.

Rowland and Molina's theory described how CFCs did not break down until they reached the stratosphere. There, ultraviolet radiation went to work on them, taking them apart and releasing chlorine atoms. These would in turn react with the ozone, leading to it breaking down as well. It was a theory dismissed by the chair of DuPont, a company among the largest producers of CFCs, as science fiction.

But the British team of Joe Farman, Brian Gardiner, and Jon Shanklin had taken Rowland and Molina's work from something theoretical to a demonstrable reality. It terrified people. Not because a scientific theory was broadly proved, but because it had an immediate consequence. The fewer ozone atoms there were in the atmosphere meant more ultraviolet radiation could pass through. And that meant skin cancer.

Public opinion coalesced around two things: fear and an image. There was "a hole in the ozone layer."

It is an image that rather stretched the reality of the science. As Nathaniel Rich put it in his staggering 2018 *New York Times Magazine* article, "Losing Earth,"* "The urgency of the alarm seemed to have everything to do with the phrase 'a hole in the ozone layer,'" which, charitably put, was a mixed metaphor. For there was no hole, and there was no layer. Ozone, which shields Earth from ultraviolet radiation, was distributed throughout the atmosphere, settling mostly in the middle stratosphere and never in a concentration higher than fifteen parts per million. But, while it may have not been the best metaphor to describe the problem, it proved ideal for embedding it into the public imagination.

What happened next would be unprecedented. The world came together,† negotiated a global treaty to deal with the problem, signed it, enacted it, and fixed it.

---

* The original article appeared on August 1, 2018, and was expanded into a book published in March 2020.

† During the Cold War, no less.

The so-called Montreal Protocol was signed in 1987 and came into effect on January 1, 1989. It bound signatories to regulate the production and consumption of Ozone Depleting Substances, and it worked. Today, the ozone "hole" is smaller than it was when scientists began to measure it in 1982.

According to the UN, in the first ten years after the Protocol was signed, greenhouse gas emissions were reduced by the equivalent of 135 gigatons of carbon dioxide. It stands as the most successful environmental agreement of all time. It also offers a road map to a similar agreement to combat carbon emissions, the main driver for the climate change that has rendered the Arctic navigable for the first time in human history. But it hasn't happened yet. And therein hangs a tale.[*]

The basic science of climate change—how the release of certain greenhouse gases, including the aforementioned CFCs and carbon dioxide, into the atmosphere as a result of human activity would raise the temperature of the planet—has been understood for decades.[†] But somehow, it managed to pass unnoticed until the late 1970s, when the then-deputy legislative director of Friends of the Earth, Rafe Pomerance, read a paper published by the US Environmental Protection Agency. He found, buried in the

---

[*]   What follows is basically a summary of Nathaniel Rich's *Losing Earth*, which remains the best telling of the story I have found.

[†]   This despite the work of several bad actors since the early 1990s to make people think otherwise.

last paragraph on a section on environmental regulation, something that shocked him: a statement that fossil fuel use would cause "significant and damaging" changes to the atmosphere.

It shocked him because he had never heard anything like it before. So he asked a colleague if she'd heard of it. She hadn't either. But she was able to point him in the direction of an essay called *How to Wreck the Environment*, written in 1968 by President Johnson's science advisor, George MacDonald. He had first studied the effects of carbon dioxide when he had worked as an advisor in the Kennedy White House seven years before.

By the time Pomerance tracked him down, MacDonald was working on yet another study on climate change, this time with the JASONs, an independent group of scientists set up in 1959 to advise the US government.

Together, they would begin a series of informal briefings to try to educate senior officials within the EPA, the National Security Council, and the Energy Department, among others, starting in the spring of 1979.

Gradually, people started to listen. And finally they scored a meeting with Frank Press, science advisor to President Carter, and his staff.

MacDonald used the briefing to take his audience through the history of the science so that they might understand how grounded it was in long-standing research.

He began with the Irish physicist John Tyndall who, in 1859, discovered that carbon dioxide absorbs heat, and that therefore higher or lower levels of it in the atmosphere could affect the climate.

This work in turn led the Nobel Prize–winning Swedish chemist Svante Arrhenius to work out that burning coal and gasoline could implement Tyndall's discovery. Arrhenius reckoned that this phenomenon would not be noticeable for centuries. But just forty years later, based on observations at weather stations across the UK, an engineer and amateur meteorologist called Guy Stewart Callendar noticed that the previous five summers, from 1934 to 1939, were the hottest in recorded history. He would go on to write that humanity was now "able to speed up the processes of Nature."

MacDonald wrapped up his potted history by telling listeners of Roger Reville, with whom he had worked under Kennedy. In 1958, Reville had helped the United States Weather Bureau begin monitoring carbon dioxide levels in the atmosphere from a station atop Mauna Loa in Hawaii at an altitude of 3,500 meters. It is this data that formed the basis of what we now know as the Keeling Curve, drawn from measurements taken daily since March 29, 1958.

That day, there were 313 parts per million of carbon dioxide in the atmosphere. Today,[*] there are 420.69 ppm.

MacDonald's briefing left no doubt that the data on carbon emissions had been known for a long time. Even

---

[*] February 24, 2022.

as far back as the 1960s, there was some call to action. On February 8, 1965, shortly after his inauguration, President Johnson told Congress that "Air pollution is no longer limited to isolated places. This generation has altered the composition of the atmosphere on a global scale through radioactive materials and a steady increase in carbon dioxide from the burning of fossil fuels."

Johnson commissioned his Scientific Advisory Committee to study carbon dioxide's effects further. With Reville as its chair, it would report that these would include melting ice caps, rising seas, and an increased acidulation of existing fresh water. It would also argue that, if we were to stop it, it would take a coordinated global plan.

Nothing happened.

At around the same time, a young NASA scientist called Jim Hansen began to study the atmosphere of Venus at the Goddard Institute of Space Studies. His goal was simple: to discover why it is so hot. The answer would come from the USSR's Venera 4 mission which, on October 18, 1967, became the first space probe to send information back from the planet. Its atmosphere was primarily made up of . . . carbon dioxide.

NASA would later expand its research to our own atmosphere, and Hansen found himself refining and applying the models he created for his Venusian research to planet Earth.

So, when Press, the MacDonald briefings no doubt ringing in his ears, wrote to Carter on May 22, 1979, to ask that the National Academy of Sciences work on the carbon dioxide issue, it was to Goddard Institute meteorologist Jule Carney, Hansen's boss, that they would turn.

Carney knew Hansen was one of the few people on earth who had modeled what carbon emissions could do to a planet. His report, *Carbon Dioxide and Climate: A Scientific Assessment*, delivered a few months later, would, in the words of Nathaniel Rich, "come to have the authority of settled fact. It was the summation of all the predictions that had come before, and it would withstand the scrutiny of the decades that followed."[*]

In fact, it was so persuasive that the oil giant Exxon would begin its own research on atmospheric carbon dioxide, primarily to find out how much blame it would have to shoulder for its share of responsibility for the looming catastrophe.

But scientists are not politicians. And, for all of the Carney report's potency, President Reagan's election in 1980 would stall all progress on the issue. In Rich's words: "Reagan appeared determined to reverse the environmental achievements of Jimmy Carter, before undoing those of Richard Nixon, Lyndon Johnson, John F. Kennedy and, if he could get away with it, Theodore Roosevelt."[†]

---

[*]   "Losing Earth," *New York Times Magazine*, August 1, 2018.

[†]   Ibid.

However, the issue would not go away, and, thanks to Congressional committee hearings chaired by then-Representative Al Gore, it would finally become a matter of public record.

Gore's interest in the environment and climate change had been triggered at Harvard, where he attended classes taught by Roger Reville. He was convinced that, once the science was known, the political class would be compelled to act. But, despite strong testimony by Jim Hansen and from Reville himself, it didn't.

It's worth stating that, in the 1980s, climate change was not the intensely partisan issue that it has become today. If anything, House Republicans, interested in conserving the planet, were the ones pressing for action. Representative Robert Walker of Pennsylvania would say: "In each of the last five years, we have been told and told and told that there is a problem with the increasing carbon dioxide in the atmosphere. We all accept that fact, and we realize that the potential consequences are certainly major in their impact on mankind . . . The research is clear. It is up to us now to summon the political will."

Six years later, in 1988, George H. W. Bush would actively campaign on the issue, saying, "I am an environmentalist," while his running mate Dan Quayle said, "The greenhouse effect is an important environmental issue. We need to get on with it. And in a George Bush administration, you can bet that we will."

It was a campaigning position galvanized by the world's reaction to the ozone "hole" revelations halfway through Reagan's second term and the success of the Montreal Protocol, a success that also prompted a bipartisan group of forty-one US senators to demand that Reagan seek an international treaty, modeled on the ozone deal, to tackle the carbon crisis.

As a result, in May 1988, Reagan and Mikhail Gorbachev, representing the world's two largest carbon emitters, signed a pledge to do just that.

But pledges are not actions. As Pomerance was coming to realize, one of the key things missing from the argument for action was something tangible that the public could latch onto. The ozone "layer" and its "hole": a visual representation that was easy to picture, with a solution that was therefore easy to imagine even if you didn't understand the science.

It raises a fundamental question for climate change politics: how do you show the public something that doesn't seem to be happening* which, when it does finally become visually apparent, we will no longer have time to solve?

This question of a visual representation of the issue is further compounded by other factors. The first of these is that we are not very good at understanding time. Our memories are remarkably short-term and centered on the experiences of people alive right now. As Angela Merkel

---

* Even though it was and is . . .

noted in a press conference given on July 20, 2018, "When the generation that survived the war is no longer here, we'll find out whether or not we learned from history." Experience suggests that we probably haven't, especially when one looks at the way some in Britain have sought to transform its national Remembrance Day now that the generations that fought two world wars are broadly gone.* The bigger the amount of time we are asked to reckon with—like the vast gaps between major earthquakes along the planet's significant fault lines, for example—the harder we find them to fathom.

Our grasp of future time is likely worse. It requires imagination and the ability to take a long view. But we live in a world dominated by the problems of the now to the exclusion of almost everything else. If it isn't going to affect us now, it is probably not going to be bothered with.

This stems from another problem that affects us as an animal: we are very bad at understanding risk. This is because our brains have two processes for assessing it. The first, from our amygdala, governs the fight or flight response. It is one of the most primitive parts of us, and most likely something our forebears evolved before they had even crawled out of the oceans. It reacts to frightening situations, like a tiger springing out of the jungle

---

* https://bylinetimes.com/2021/11/11/a-war-christmas-what-exactly
  -are-we-remembering/.

when you're picking mangoes,* flooding your system with the adrenaline required to help you react quickly. But, as anyone who has ever suffered from anxiety will tell you, it is fundamentally flawed and can also trigger this response to threats that don't even exist. It is not rational. It can actively harm us, provoking rage and fear when they ought not to exist.

The second process is based in our neocortex, a comparatively new part of the brain.† It is analytic, nuanced, and slow.

These two work side by side. But the primitive amygdala frequently trumps the modern neocortex, taking reason away from our risk analysis. Thus, if the risk before you is not a tiger but, let's say, a potential investment, should something happen to fire the amygdala while the neocortex is doing its thing, your understanding of that risk will be flawed.

We also suffer here with a problem of imagination. We can imagine the tiger and the investment, but we are bad at imagining the effects we have on others or on the world around us. As Marie says to Jess in *When Harry Met Sally*,‡ as they argue over his wagon wheel coffee table: "Everyone in the world thinks they have a sense of humor and good taste but they don't all—" We perceive ourselves

---

* See chapter 1: "Never get off the boat."

† It exists only in mammals.

‡ Scene 78.

as we would like to be, not as we are. We are all the heroes of our own stories. And we find it impossible how little old us could possibly affect an entire planet because we think of it on individual terms. We don't, can't, imagine "us" as the 7.9 billion of us there really are.

There is another problem. We've been talking about climate change since the 1980s. We are overexposed to it. It has not suddenly sprung upon us like a tiger or a lion, nor like the "hole" in the ozone layer.

For all these quirks of human nature that stand in our way as we grapple with climate change in the twenty-first century, back at the beginning of 1989, there was a real reason for optimism that something would get done. The Montreal Protocol offered a template for a similar global treaty devoted to carbon emissions. The goal of a 20% reduction by 2000 was on the table. The new president George H. W. Bush had, as we've seen, declared himself an environmentalist and had run on the issue. In an America that would have to lead the charge, as it had done on CFCs, there was a broad bipartisan consensus in Congress that this was a good and right thing to do.

That April, as Jim Hansen prepared to testify before Al Gore's latest hearings on the subject, EPA administrator William Reilly went to the White House with a new proposal: that, at the meeting of the Intergovernmental Panel on Climate Change meeting in Geneva that May and at an international meeting to take place in the

Netherlands that autumn, the US should demand a global treaty to tackle carbon dioxide emissions.

His proposal would be shot down by John Sununu, the new president's Chief of Staff.

Sununu, it would seem, was an early adopter of something the world has become all too familiar with in recent years: he did not believe the science. And his boss, the president, was uninterested. As Rich writes: "When Reilly tried in person to persuade him to take action, Bush deferred to Sununu and [former Secretary of State James] Baker. Why don't the three of you work it out, he said. Let me know what you decide."[*]

Sununu ordered the US delegates at the IPCC meeting in Geneva to make no commitments on carbon emissions. And it was about to get worse.

On November 6, 1989, ministers from sixty-eight countries sat down at what would become known as the Noordwijk Climate Conference, the first such meeting of its kind. Their goal was to create a binding agreement on carbon dioxide emissions that would form the basis of a global treaty similar to the Montreal Protocol.

Although not one of the official delegates, Pomerance went along—he had a vested interest, after all—together with three colleagues, Daniel Becker, Alden Meyer, and Stewart Boyle. Their aim: to keep the pressure on the

---

[*] "Losing Earth," *New York Times Magazine*, August 1, 2018.

delegates to commit the agreement's signatories to a 20% emissions reduction by 2005.

Between them, they kept the story firmly in front of the media, but to no avail.

Even though most ministers present had come to the conference ready to sign the agreement, negotiations continued long into the night, with Pomerance and his colleagues starved of news about their content outside.*

When Becker collared the Swedish minister, who had slipped out for a bathroom break, about what was happening, he was told: "Your government is fucking this thing up!"

Sure enough, the American delegate Allan Bromley, with the support of delegates from Britain, the Soviet Union, and Japan, had managed to force the conference into inaction.

The gig was up.

In the thirty-three years since, the American political consensus on climate change would be gradually dismembered. Many nations have failed to meet the commitments they made in Geneva and Noordwijk. The oil industry has shifted its position from at least trying to understand the problem to employing tactics akin to the tobacco industry in its attempts to avoid regulation and to obfuscate debate, forming a lobby group called the Global Climate Coalition, which would go on to launch a $13 million advertising campaign to undermine support for 1997's Kyoto Protocol.

---

* "Losing Earth," *New York Times Magazine*, August 1, 2018.

Kyoto committed signatories to just a 5% reduction, taking them back to 1990 levels. The US Senate voted 95–0 to express its opposition to ratifying the protocol.

We find ourselves living in a time when political leaders, particularly in the United States, view climate science through a prism of ideology rather than pragmatism.

In an article for the European Journal of American Studies,* Jean-Daniel Collomb writes, "There is an ideological dimension to the effort to counter climate action: the conservative movement appears to be committed to small government and free enterprise as ideological ends in themselves, irrespective of economic and environmental common sense. From the small-government perspective, therefore, discrediting calls for strong national and international climate action has become a matter of ideological survival."

Later in the article, he goes further: "Global warming poses a philosophical challenge to libertarians and small-government conservatives: their world view is premised on the idea that government power should always be held in check lest it destroy individual freedom while the world is faced with a crisis of global proportions that could only be averted by a strong and prolonged government action," before stating: "Denial appears to be a more desirable strategy than a devastating reappraisal of one's deeply held beliefs." Which makes me want to refer them to the famous

---

* *The Ideology of Climate Change Denial in the United States*, Spring 2014.

wager posited by the French philosopher Blaise Pascal,* frequently expressed as a quote attributed to Albert Camus:† "I would rather live my life as if there is a god and die to find out there isn't, than live my life as if there isn't and die to find out there is."

Such ideology cannot be challenged by public apathy. But, as Jeffrey Kluger wrote in an article for *Time*,‡

> The bad news for environmental scientists and policymakers trying to wake the public up to the perils we face is that climate change checks almost every one of our ignore-the-problem boxes. For starters, it lacks the absolutely critical component—the "me" component. "Nobody wakes up in the morning and looks at the long-term climate forecast," says David Ropeik, an international consultant on risk perception and communication, formerly with the Harvard School of Public Health. "They ask what the weather is today, where I live, and how it's going to affect me."

---

* Assuming they'd do anything as radical as reading something by a Frenchman.

† And you know how good the internet is at mistakenly attributing quotations.

‡ October 8, 2018.

Later in the same piece, Kluger quotes Paul Slovic, a psychologist at the University of Oregon, who says, "If we think the consequences are far in the future, we tend to discount the risk. People just aren't going to inconvenience themselves unless they're forced to." Slovic continues, "The question is often, 'Do I feel vulnerable?' For the most part we don't and that shapes our behavior."

All of this brings us back to the thing the ozone crisis had and which climate change does not: an image to embed its urgency in the public mind. A photographer called James Balog may have found it.

Balog, the subject of the award-winning 2012 documentary *Chasing Ice*, has stated that, as little as ten years before he began to attract public attention, he had been a climate change skeptic. As he put it in a TED talk in July 2009, "I thought that the story of climate change was based on computer models. I hadn't realized it was based on concrete measurements of what the Paleo climates were, where the ancient climates were as recorded in the ice sheets, as recorded in deep ocean sediments as recorded in lake sediments, tree rings and a lot of other ways of measuring temperature. When I realized that climate change was real, and it was not based on computer models, I decided that one day I would do a project looking at trying to manifest climate change photographically."

It would prove a project for which, as a nature photographer, he was extremely qualified. And it would be conceived on a stills shoot for *National Geographic* in 2006, photographing glaciers for what would become a cover story called "The Big Thaw."* "I got the idea that I should shoot in time-lapse photography, that I should station a camera or two at a glacier and let it shoot every 15 minutes or every hour or whatever, and watch the progression of the landscape over time."†

The Extreme Ice Survey was born. Its mission: to place time-lapse cameras where they could photograph glaciers in Alaska, the Rockies, Greenland, Iceland, and the Alps,‡ shooting a frame every hour, year-round, in daylight. It yields some 8,000 shots per camera per year and, when put together in animation, reveals dramatic footage of climate change in action. Each of these glaciers is thinning and retreating on a shocking scale.

As Balog puts it, "These images make the invisible visible." In *Chasing Ice*, he says, "If I hadn't seen these pictures, I wouldn't have believed it at all." Nor that it is happening so fast.

Balog's diagnosis is that we have a perception problem surrounding climate change. "If you had an abscess in your tooth, would you keep going to dentist after dentist until

---

* June 2007.

† This quote, again, comes from his July 2009 TED talk.

‡ They also operate two sites, one in the Andes and one in Canada, for repeat visits rather than permanent positions, and they have recently begun work in Antarctica.

you found a dentist who said, 'Ah, don't worry about it. Leave that rotten tooth in'? Or would you pull it out because more of the other dentists told you that you had a problem? That's sort of what we're doing with climate change.'" And he thinks part of that stems from the fact that people tend to think of geology as something that happened a long time ago, rather than something that is happening right now.

This feeds into our human trouble with understanding large chunks of time, an inability which makes us stupid.

In 2015, Katherine Schultz wrote an article for *The New Yorker*† that illustrates this well. Entitled "The Really Big One," it details a geologic fault line that is largely unknown of by most of us, the Cascadia subduction zone. It runs beneath the ocean from Vancouver Island in the north to Northern California in the south. Nobody can remember the last time it caused a major earthquake. When it does, it will likely trigger a tsunami that will wipe out the Pacific Northwest.

Its last big performance took place on January 26, 1700. That we know about it at all stems from the linking of a strange account of a rogue tsunami, with seemingly no source, that hit the eastern coast of Japan on January 27, 1700, to Native American accounts from the Pacific

---

*   *Chasing Ice.*

†   July 20, 2015.

Northwest that told of a tsunami that wiped out entire tribes.*

Schultz writes, "We now know that the Pacific Northwest has experienced forty-one subduction-zone earthquakes in the past ten thousand years. If you divide ten thousand by forty-one, you get two hundred and forty-three, which is Cascadia's recurrence interval: the average amount of time that elapses between earthquakes. That timespan is dangerous both because it is too long—long enough for us to unwittingly build an entire civilisation on top of our continent's worst fault line—and because it is not long enough. Counting from the earthquake of 1700, we are now three hundred and fifteen years into a two-hundred-and-forty-three-year cycle."

We are overdue. We don't, unless we're experts, understand these kinds of timescales.

And, as we've seen, we no longer have the generational memory to tell us what we don't know.

Balog often calls glaciers "the canary in the climate change coal mine."† We need to wake up to the consequences of their vanishing.

---

* From the Japanese end, we have to thank the seismologist Kenji Satake; from the Native American side, Chief Louis Nookmis, of the Huu-ay-aht First Nation from British Columbia, who recounted a seven-generation old story about the eradication of Vancouver Island's Pachena Bay people; and a 2005 study by Ruth Ludwin and a team at the University of Washington which catalogued and analysed Native American oral stories of earthquakes and tsunamis.

† He does so in both his 2009 TED talk and in *Chasing Ice*.

On February 7, 2021, a part of a Himalayan glacier broke away, sending a surge of water down the Dhauliganga River in the Chamoli district of Uttarakhand in Northern India. It destroyed roads, bridges, and two hydroelectric plants, and, as the news broke the following day, had left twenty-six people dead and at least 170 missing.*

As tragic as such loss of life is, and as likely as such events are to happen again, there is a bigger picture to consider. The Himalayas have the world's third largest repository of snow and ice, and their glaciers supply the Ganges, the Brahmaputra, the Indus, the Mekong, the Yangtze, and the Yellow rivers, providing for the water needs of one-and-a-half billion people. Even a small percentage drop in their flow would spell catastrophe.

But it is the melting of the Thwaites Glacier in Antarctica that is perhaps our biggest cause for concern. Estimated by some to be of equivalent size to the state of Florida, it holds back the entire West Antarctic Ice Sheet (WAIS). It contains enough ice to, if melted, raise sea levels by three meters. The impact of that upon low lying countries like Bangladesh, Qatar, or the Netherlands would be catastrophic. The Maldives would be wiped from the face of the earth. Large swathes of New York City, New Orleans, and Florida would find themselves

---

* https://www.sbs.com.au/news/article/at-least-26-people-dead
  -170-missing-in-northern-india-after-huge-chunk-of-himalayan
  -glacier-crashes-into-river/1q62d6e69.

underwater.* In fact, some 250 million people live within a meter of high tide marks around the world.

We know from the work of James Balog and others that, even though we cannot see it with the naked eye, big changes can strike a melting glacier incredibly fast. And they may prove irreversible.

In the case of the Thwaites, that would be very bad news for us indeed. Not for nothing has it been dubbed by the journalist Jeff Goodell "The Doomsday Glacier."[†]

The Thwaites is what is known as a "threshold system." This means that it is one whose success or failure will fundamentally switch the things related to it from success to failure as well. In the Thwaites's case, that means it will hold . . . until it doesn't. And what the scientists working on the International Thwaites Glacier Collaboration are trying to determine is how stable is it, and when might be the point of no return.

In Goodell's most recent piece on the Thwaites,[‡] he writes, "The west Antarctic ice sheet [bottled up by the Thwaites[§]] is one of the most important tipping points in the Earth's climate system," before going on to quote Ian Howat, a glaciologist at Ohio State University, saying, "If

---

*   To add a little perspective, NASA analysis shows that sea levels have risen just 20 cm between 1901 and 2018.

†   *Rolling Stone*, May 9, 2017.

‡   *Rolling Stone*, December 28, 2021.

§   My words, added for clarity.

there is going to be a climate catastrophe, it is probably going to start at Thwaites."

Why it is of such concern is this: unlike other glaciers, which are melting thanks to contact with warmer air than normal, the Thwaites is melting from below. Warmer ocean water is eating away at its base. This matters because the ice sheet behind it sits upon what Sridhar Anandakrishnan, a glaciologist at Penn State University, describes as a giant soup bowl.

Writes Goodell: "What this means is that once the warm water gets below ice, it can flow down the slope of the bowl, weakening the ice from below. Through a mechanism called 'marine ice-cliff instability,' you can get what amounts to a runaway collapse of the ice sheet that could raise global sea levels very high, very fast."*

As Rob Partner, a geophysicist with the British Antarctic Survey says in the same article, "The net rate of ice loss from Thwaites Glacier is more than six times what it was in the early 1990s."†

Or, to put it another way, change is happening, fast, right now. And it does not look good. Especially when you realize that the Antarctic has lost some three trillion tons of ice in the last twenty-five years.

---

* Ibid.

† Ibid.

This brings with it yet another, almost unimaginable problem. As the materials scientist Mark Miodownik wrote recently, the Earth isn't solid:

> Right at the center of the Earth is a solid core of metal made of iron and nickel at a temperature of approximately 5,000°C.* But this core is surrounded by an approximately 2,000 km-thick ocean of molten metal, again mostly iron and nickel. Surrounding this is a layer of rock called the mantle that is between 500°C to 900°C,† and at these red-hot temperatures the rock behaves like a solid over short periods of time (seconds, hours, and days) but like a liquid over longer time periods (months to years)—so the rock flows, even though it is not molten. On top of the fluid mantle floats the crust, which is like the skin of the Earth.‡

Both Antarctica and Greenland are covered in ice up to depths of as much as three kilometers in places, creating significant weight that pushes those landmasses down into the fluid mantle. Were that ice to disappear, it would not only raise sea levels around the world, but that weight

---

* 9,032°F.

† Approximately 932–1,652°F.

‡ The *Guardian*, October 1, 2018.

would be gone. And without it, the land above, floating as it is, will bob up.

This is called "post-glacial rebound." Its consequences are not fully understood. Miodownik writes, "we badly need more data."* If the ice in Greenland should melt first, then the North American tectonic plate will rise. But by how much? And if Antarctica should be first, the northeastern seaboard of the United States will be submerged. "The big unknowns are how quick the ice will go in each location, and how fast the post-glacial rebound will be," Miodownik says.†

Perhaps more worryingly, this is a climate change issue in which some countries might win territory and some might lose. This further politicizes an issue which has been riddled with political malaise and inaction since the Kyoto Protocol of 1997 and the Paris Climate Accords of 2015.

If carbon dioxide emissions, melting ice, the loss of key water supplies, and post-glacial rebound weren't enough for you, there is yet another contributor to this all-out assault on civilization as we know it: methane.

Methane is a hydrocarbon found in natural gas produced by bacteria living in anaerobic environments like bogs, swamps, or other wetlands. It's also a by-product of coal mining, where ancient gas can be released frequently as miners cut into the coal seams. It makes

---

*   Ibid.

†   Ibid.

up between 17–20% of global greenhouse emissions. This may seem like small potatoes compared to carbon dioxide, which makes up 75% of the total. But methane is twenty-eight times better than CO_2 at trapping heat in our atmosphere. In fact, UN experts estimates that, over the course of twenty years, the number shifts upwards, becoming more than eighty times more efficient.

The good news is that methane in the atmosphere breaks down in roughly ten years, compared to carbon dioxide's three hundred. Which means that, if the world gets its act together, tackling methane emissions will be a substantial step towards meeting the commitments made in the Paris Climate Accords to keep the rise in mean global temperature under 2°C (35.5°F).

However, at both Kyoto and Paris, this very solvable issue never came up. And it wasn't on the agenda for the COP26 Conference held in Glasgow in 2021 either.

And yet, at that Conference, one hundred countries, including two of the biggest methane emitters, the United States and Brazil, signed up for a 30% reduction by 2030. It was one of the Conference's most significant achievements. In his address to the Conference, US president Joe Biden said, "I think we can probably go beyond that."[*]

---

[*]   Dominic Faulder, *Methane, The Overlooked Greenhouse Gas, Nikkei Asia*, November 8, 2021.

But how?

Remarkably easily.

Methane emissions are all connected to human activity. It seeps out of current and abandoned coal mines.* It leaks from badly maintained gas pipes. It escapes from Arctic and sub-Arctic tundra as the planet warms. It is vented by energy companies. And it is produced by agriculture—cattle and rice farming.

Admittedly, the tundra issue remains a challenge. But leaky gas pipes can be fixed. Venting can be banned. Abandoned coal mines can be capped. The requisite farming practices can be changed. And we know how.

The significant issue remaining is how to bring the other three of the top five methane emitters, Russia, India, and, largest of all, China, along with the hundred signatories to the methane agreement.

And here we come up against another obstacle. If you ask most people, "Do you want to protect the environment?" the answer is usually "yes." If you follow that up with, "At what cost to yourself?" the subtext of their reply usually boils down to, "As little as possible."

But everything comes at a cost.

The cost of action affects us all. In the short term.

The cost of inaction is incalculable. And forever.

---

*   According to the IEA, the methane equivalent of almost 3.5 billion tons of carbon dioxide leaked from coal mines in 2020.

Because this is not about saving the planet. Planet Earth will carry on doing what it does, orbiting the sun every 365-and-a-quarter days. It will go on. We—our civilization—won't.

There's a line in the film *The Killing Fields*, in which the journalist Sydney Schanberg is trying to arrange a passage out of Cambodia for his college Dith Pran with the US Consul, played by Spaulding Grey, as he supervises the shredding of everything in the embassy. As Schanberg expresses his outrage that the US will not even be airlifting in much needed food once they've gone, the Consul justifies it, saying, "Excuse the pun, but we're either staying or we're living."

That's where we are right now. Globally. We're shredding documents in an embassy before the enemy arrives and kills us all, if only we can get out first.

Except there is no "out." There is nowhere to go. So we must bear the cost of action, and we must do it now.

As we saw at the beginning of this book, civilization itself came at a cost. Ice, as a recent gift to civilization, on the scale we know it, comes with one, too.

That cost may just be civilization itself.

ACKNOWLEDGMENTS

B ooks don't happen in a vacuum. For all the trudging
through libraries, straining the eyes at a microfiche,
reading, typing, and swearing that goes into writing a book
like this, there is still a vital team who take all that and
turn it into the volume you hold in your hand. My thanks,
firstly, to my agent Charlie Brotherstone for embracing
all these mad tales I presented him with and steadfastly
pushing me to turn them into this. Thanks equally to
my publisher, Claiborne Hancock, who, in addition
to being a pleasure to work with, has gracefully put up with
a twelve-hour time difference every time we've spoken on
the telephone. I am grateful, too, to copy editor Victoria
Flickinger, who has not only made me look like a much
better writer than I am, but has faced the mountainous
task of turning each of my British spellings into American
Standard English. And to Maria Fernandez, who has han-
dled all the production side of bringing this book to press,
a job in which she has not been helped by me blowing my
deadline. My thanks, too, to Jessica Case, who has not
only coordinated all conversations between New York and

Thailand, but is also in charge of the matter of publicity, without which you probably would never read this. The fine team at Faceout Studios also deserve a shout-out for their cover design—I fell in love with it at first sight, and not just because it seemed so inspired by one of the Winter Olympics posters from the 1972 Sapporo Games. And I'd also like to thank publicity guru Nicole Maher for going to the mats for this project. I feel most fortunate to have such a team on my side.

I've also been very lucky to have had a lot of support from friends while I've been working on this, sadly too many to name here. To single out just two, both Jonathan Wakeham and Tania Kindersley have listened to me ad nauseam as I've run arguments past them and pondered ideas, and both have made excellent suggestions to help me improve the text. Thank you.

And, of course, my family: my father, Philip; my brother, Gavin, and sister-in-law, Ann Marie; and mostly, my wife, Karen, who has read everything I've typed from first outlines to final proofs, and has been endlessly encouraging. My love, this could not have been written without you.

BIBLIOGRAPHY

Albert, Thomas F. "The Influence of Harry Brower, Sr., an Iñupiaq Eskimo Hunter, on the Bowhead Whale Research Program Conducted at the UIC-NARL Facility by the North Slope Borough." Arctic Institute of North America, 2000.

"Arctic Exploration," *The Canadian Encyclopedia*, 2017. https://www.thecanadianencyclopedia.ca/en/timeline /arctic-exploration.

"Arctic Exploration," *The Canadian Encyclopedia*, January 3, 2020. https://www.thecanadianencyclopedia.ca/en/article /arctic-exploration-editorial.

Arthur, Stanley Clisby. *Famous New Orleans Drinks and How to Mix Them.* Pelican Publishing Company: Gretna, Louisiana, 1937.

Beattie, Owen, and Geiger, John. *Frozen In Time: The Fate of The Franklin Expedition.* London: Bloomsbury, 1987.

Beaudry, Frederic. "Ski Resorts and Their Impact on the Environment," Treehugger.com, November 3, 2019.

Becker, Gary S. "A Theory of the Allocation of Time," *Economic Journal* 75, 299 (1965): 493–517.

Belon, Pierre. *Les Observations de Plusieurs Singulaitez et Choses Memorables en Grece.* Paris: Asie, 1553.

Brewster, Karen, ed. *The Whales They Give Themselves.* Fairbanks: University of Alaska Press, 2004.

Buxbaum, Tim. *Icehouses.* Oxford: Shire Publications, 2014.

Collomb, Jean-Daniel. "The Ideology of Climate Change Denial in the United States," *European Journal of American Studies*, Spring 2014.

Dahl, Jacob (project director). *The Electronic Text Corpus of Sumerian Literature Faculty of Oriental Studies*. Oxford: University of Oxford, 2003, 2004, 2005, 2006.

David, Elizabeth. *Harvest Of the Cold Months*. London: Michael Joseph, 1994.

Denning, Andrew. "How Skiing Went from The Alps to The Masses," *The Atlantic*, February 23, 2015.

DeWall, Richard A. "The Origins of Open Heart Surgery at the University of Minnesota 1951 to 1956," *Reflections of the Pioneers*, Vol. 142, Issue 2, August 1, 2011: 267–269.

Dolan, Eric Jay. *When America First Met China*, New York City: Liveright Publishing Corporation, 2012.

Dresbeck, LeRoy J. "The Ski: Its History and Historiography," *Technology and Culture*, Vol. 8, No. 4 (October 1967): pages 467–479.

Easterlin, Richard A. *Birth and Fortune: The Impact of Numbers on Personal Welfare*, New York: Basic Books, 1980.

Easterlin, Richard A. *Population, Labor Force, and Long Swings in Economic Growth: The American Experience.* New York: Columbia University Press, 1968.

Eco, Umberto. "How The Bean Saved Civilisation," *New York Times*, April 18, 1999.

Eco, Umberto. *Six Walks in The Fictional Woods.* Cambridge, Massachusetts: Harvard University Press, 1994.

Engber, Daniel. "Who Made the Ski Lift," *New York Times Magazine*, February 23, 2014.

English, Otto. "A War Christmas: What Exactly Are We Remembering," *Times*, November 11, 2021. https://by linetimes.com/2021/11/11/a-war-christmas-what-exactly-are-we-remembering/.

Ercan, Asli. "From Mother-Goddess to the Pandora's Box: Glass Ceiling Myth." *European Journal of Research on Education*, July 2, 2014.

Evelyn, John. *The Diary of John Evelyn*. Oxford: Oxford University Press, 1959.

Faulder, Dominic. "Methane, the Overlooked Greenhouse Gas," *Nikkei Asia*, November 8, 2021.

Gardien, Claude. "Inaccessible," *The Alpinist*, September 14, 2017.

Gawande, Atul. *The Checklist Manifesto*. London: Profile Books, 2010.

Gallant, Joseph. "The World Comes Together in Your Living Room: The Olympics on TV," *TV Broadcasting History*. https://jeff560.tripod.com/tv8.html.

Goldman, William. *Which Lie Did I Tell?* New York City: Pantheon Books, 2000.

Goodell, Jeff. "The Doomsday Glacier". *Rolling Stone*, May 9, 2017.

Goodell, Jeff. "'The Fuse Has Been Blown,' and the Doomsday Glacier Is Coming for Us All," *Rolling Stone*, December 28, 2021.

Healey, Nathalie. "Extreme Cold Is Bringing Humans Back from the Brink of Death," *WIRED*, February 5, 2020.

Henderson, Alexander. *The History of Ancient and Modern Wines*. London: Baldwin, Cradook, and Joy, 1824.

Henson, Mathew A. *A Negro Explorer at The North Pole*. New York: Frederick A. Stokes Company, 1912.

Hickman, Leo. "Can You Ski and Be Green?" *Guardian*, September 28, 2008.

Ingels, Margaret. *Willis Carrier: Father of Air Conditioning*. New York: Country Life Press, 1952.

Jackson, Tom. *Chilled*. London: Bloomsbury, 2015.

Johnson, Steven. *How We Got to Now*. London: Particular Books, 2014.

Konstantinov, Igor E. "Vasilii I. Kolesov: A Surgeon to Remember," *Texas Heart Institute Journal*, 31(4), 2004: 349–358.

Kovarik, William. "Ethyl-leaded gasoline: How a classic occupational disease became an international public health disaster," *International Journal of Occupational and Environmental Health*, 11 (4), 2005.

Kozlov, Max. "Doubts Raised About Cooling Treatment for Oxygen-Deprived Newborns," *Nature*, August 11, 2021.

Maugh II, Thomas H. "Wilfred G. Bigelow, 91; Cardiac Specialist Invented Pacemaker," *Los Angeles Times*, April 1, 2005.

Miller, Gifford H. and Áslaug Geirsdóttir, Yafang Zhong, Darren J. Larsen, Bette L. Otto-Bliesner, Marika M. Holland, David A. Bailey, Kurt A. Refsnider, Scott J. Lehman, John R. Southon, Chance Anderson, Helgi Björnsson, Thorvaldur Thordarson. "Abrupt Onset of the Little Ice Age Triggered by Volcanism and Sustained by Sea-Ice/Ocean Feedbacks," *Geophysical Research Letters*, vol. 39, L02708, 2012.

Miodownik, Mark. "Which Cities Will Sink into The Sea First? Maybe Not the Ones You Expect," *Guardian*, October 1, 2018.

Nagengast, Bernard. "It's a Cool Story," *Mechanical Engineering*: 122.

Novak, Matt. "The Great Depression and the Rise of the Refrigerator," *Pacific Standard*, October 9, 2012, updated June 14, 2017.

Oster, Matt. "The History of The Weekend," *Hankering for History*, 2014, https://hankeringforhistory.com/history-weekend/.

Palin, Michael. *Erebus: The Story of a Ship*. London: Hutchinson, 2018.

Pendrous, Rick. "Fridges Heralded the UK's Chilled Food Chain," *Food Manufacturer*, August 24, 2017. https://www.foodmanufacture.co.uk/Article/2017/08/25/Fridges-heralded-the-UK-s-chilled-food-chain.

Persall, Frank V. "Who Invented the Ski Lift? An Engineering Masterpiece," Snowgaper.com.

Pollack, Henry. *A World Without Ice*. London: Penguin, 2012.

Potter, Russell. "Remembering Louie Kamookak," *Nunatsiaq News*, April 2, 2018.

Project Jukebox, University of Alaska Fairbanks Oral History Program, interview conducted by Bill Schneider and David Krupa, October 29, 2002.

Rayner, Jay. "Brexit Provides the Perfect Ingredients for a National Food Crisis," *Guardian*, July 29, 2018.

Rees, Jonathan. *Refrigeration Nation*. Baltimore: The John Hopkins University Press, 2013.

Rich, Nathaniel. "Losing Earth," *New York Times Magazine*, August 1, 2018.

Rimmer, MBChB, Lara. Matthew Fok, MBChB, and Mohamad Bashir, MD, MRCS. "The History of Deep Hypothermic Circulatory Arrest in Thoracic Aortic Surgery," *Aorta*, Vol. 2, Issue 4, June 2014: 129–134.

Ross, Ellie. "Hot Topic: Can Skiing Ever Be Green?" *National Geographic*, April 8, 2019.

Scoresby Jr., William. *Journal of a Voyage to the Northern Whale Fishery*. Edinburgh: Archibald Constable Company, 1823.

Silver, Carly. "Do You Want to Build an Ice House?" *Lapham's Quarterly*, September 2021.

Stannard, David. *Fishing Up the Moon: Norfolk Seafood Cookery*. Norwich: Larks Press, 2005.

The Editorial Team. "Venta Maersk Calls at St. Petersburg Concluding Passage of Northern Sea Route." Safety4Sea. October 1 2018. https://safety4sea.com/venta-maersk -calls-at-st-petersburg-concluding-passage-of-northern -sea-route/.

Thomas, Jerry. *The Bar-Tender's Guide*, New York: Dick and Fitzgerald, 1862.

Thompson, Hunter S. *The Great Shark Hunt: Gonzo Papers Vol. 1.* New York: Simon & Schuster, 1979.

Thomson, Helen. "Exclusive: Humans Placed in Suspended Animation for the First Time," *New Scientist*, November 20, 2019.

Tisherman, Samuel A. "Emergency Preservation and Resuscitation for Cardiac Arrest from Trauma," *Annals of the New York Academy of Sciences*, Vol. 1509, Issue 1, March 2022.

Tisherman, Samuel A. "Therapeutic Hypothermia in Severe Trauma," *ICU Management & Practice*, Vol. 13, Issue 4, Winter 2013/2014.

"At Least 26 People Dead, 170 Missing in Northern India after Huge Chunk of Himalayan Glacier Crashes into River," *SBS News*, February 8, 2021. https://www.sbs .com.au/news/article/at-least-26-people-dead-170 -missing-in-northern-india-after-huge-chunk-of -himalayan-glacier-crashes-into-river/1q62d6e69.

United Kingdom Food Security Report 2021: Theme 2: UK Food Supply Sources, updated December 22, 2021. https://www.gov.uk/government/statistics/united-king dom-food-security-report-2021/united-kingdom-food -security-report-2021-theme-2-uk-food-supply-sources.

Venzke, E., ed., "Global Volcanism Program, 2013. Volcanoes of the World." *Smithsonian Institution*, vol. 4.10.3. Accessed December 9, 2021. https://doi.org/10.5479 /si.GVP.VOTW4-2013.

Visher, S. S. "Francois Emile Matthes, 1874–1948." *Annals of the Association of American Geographers*, vol. 38, no. 4, [Association of American Geographers, Taylor & Francis, Ltd.], 1948, pp. 301–04, http://www.jstor.org /stable/2560919.

Watson, Paul. *Ice Ghosts: The Epic Hunt for the Lost Franklin Expedition*, New York: W.W. Norton & Co., 2017.

Weightman, Gavin. *The Frozen Water Trade*. London: Harper Collins, 2002.

Whitely, Michael. "How Was Ice Cream Created Before Refrigerators Were Invented," *Guardian*. https://www .theguardian.com/notesandqueries/query/0,,-1591,00.html.

Wood, Michael. *Legacy: The Origins of Civilization*. Central Television, 1991.

Younes, Osama, Reham Amer, Hosam Fawzy, and Gamal Shama. "Psychiatric Disturbances in Patients Undergoing Open-Heart Surgery," *Middle East Current Psychiatry*, September 17, 2019.

Zerubavel, Eviatar. *The Seven Day Circle: The History and Meaning of the Week*. Chicago: University of Chicago Press, 1989.

INDEX